T0275545

Ozone Crisis

The Wiley Science Editions

Ozone Crisis
The 15-Year Evolution of a Sudden Global Emergency

SHARON ROAN

WILEY SCIENCE EDITIONS

WILEY

John Wiley & Sons, Inc.

New York □ Chichester □ Brisbane □ Toronto □ Singapore

Publisher: Stephen Kippur
Editor: David Sobel
Managing Editor: Frank Grazioli
Interior Design and Production: WordCrafters Editorial Services, Inc.

Text Credits

page 29: Quotation from Angela Singer interview with James Lovelock, "Danger Theory in the Can," December 18, 1974, reprinted by courtesy of the *Yorkshire Post*; page 111: Quotation from James Lovelock, "Causes and Effects of Changes in Stratospheric Ozone: Update 1983." *Environment*, Vo. 26, No. 10, p. 26, December 1984. Reprinted with permission of the Helen Dwight Reid Educational Foundation. Published by Heldref Publications, 4000 Albemarle St., N.W., Washington, D.C. 20016. Copyright © 1984; page 253: Quotation from Wallace S. Broecker, "Unpleasant Surprises in the Greenhouse?" Reprinted by permission from *Nature*, Vol. 328, p. 123. Copyright © 1987 Macmillan Journals Limited.

This publication is designed to provide accurate and authoritative information in regard to the subject matter covered. It is sold with the understanding that the publisher is not engaged in rendering legal, accounting, or other professional service. If legal advice or other expert assistance is required, the services of a competent professional person should be sought. FROM A DECLARATION OF PRINCIPLES JOINTLY ADOPTED BY A COMMITTEE OF THE AMERICAN BAR ASSOCIATION AND A COMMITTEE OF PUBLISHERS.

Library of Congress Cataloging-in-Publication Data

Roan, Sharon.
  Ozone crisis: the 15-year evolution of a sudden global emergency.

  (Wiley science editions)
  Bibliography: p.
  1. Atmospheric ozone—Environmental aspects.
I. Title. II. Series.
TD885.5.085R63 1989   363.7'392   88-33952
ISBN 978-1-62045-637-8

90 91 10 9 8 7 6 5 4 3 2 1

Foreword

When I began my campaign for the 1988 Democratic nomination, I said that the depletion of the ozone layer, the greenhouse effect, and the global environmental crisis were the most important issues this country would have to face in the next decade and the next century. The political pundits, to put it mildly, were not impressed. Some said that anyone who thought *ozone* belonged on the next president's agenda was not going to become the next president.

The pundits were right about that, as it turned out, but they were wrong about the issue. After years of neglect, global warming and stratospheric ozone depletion have suddenly become matters of widespread international concern. The escalation of the ozone crisis even jump-started an unprecedented international treaty. While we have hardly begun to make the dramatic changes that will be necessary, it is a revolutionary step for society to recognize the problem at all.

For those who first raised these issues, the sudden awakening of public concern about the global environmental crisis may seem a triumph. At last, people will no longer dismiss the destruction of the earth's stratospheric ozone shield as the stuff of bad science fiction. But the real work now begins. This enormous crisis, which has been staring us in the face for more than a decade, poses daunting challenges that remain to be solved.

Even now, there are those who look for every possible excuse not to act, clinging to the frail human hope that perhaps these problems will somehow go away. Some want to keep waiting for more evidence, for an even clearer picture of the dire consequences we face. Others seem to think it would be easier to *adapt* to a crisis than prevent it—in the view of one Reagan administration official, to deal with holes in the ozone by passing out sunglasses and floppy hats. Almost everyone is overwhelmed by the prospect of solutions that will be harder than anything we have done before.

So even though we can rejoice at some progress—that, for example, the Congress is now considering a ban on the production of ozone-eating chlorofluorocarbons (a step that seemed unthinkable just a few years back)—the path ahead is still unimaginably difficult. We know what to do, but we do not yet know whether we have the courage within ourselves to do it.

This book will help us confront that central question. In *Ozone Crisis*, Sharon Roan recounts the journey of an environmental issue from the laboratory to the international negotiating table. The book profiles the brilliant scientists who saw trouble looming and did everything they could to show the world what it meant. It is a revealing portrait of a society with its head in the sand—industry, politicians, and the general public stubbornly refusing to accept change.

I hope we can learn from this disturbing saga. It took us 15 years to see what we have done to the atmosphere. We cannot afford to wait another 15 years to clean it up.

Senator Al Gore

February 1989

Preface

In May 1990, the United States government rejected a proposal by international leaders that would lead to protection of the earth's life-sustaining ozone layer.

Rejected it because it would cost too much money.

Thus, 15 years after the first warning, the world is not much closer to the elimination of manmade chlorofluorocarbon gases that threaten to destroy the ozone layer.

Ozone is a pungent, poisonous gas. Wafting some 15 miles above the earth's surface, it helps shield living things from the sun's searing ultraviolet light. If ozone is lost, more ultraviolet light reaches the earth, which can lead to increasing rates of skin cancer, the death of some microorganisms, and the failure of crops and plants. It was in 1974 that University of California, Irvine, chemists F. Sherwood Rowland and Mario Molina discovered that chlorofluorocarbons—the ubiquitous household chemicals that comprise a billion-dollar world-wide industry—could rise slowly to the upper atmosphere and destroy the earth's fragile ozone shield.

Rowland and Molina naively hurried to alert the world of their findings. Although their theory was judged to be correct in 1976, they were to discover that billions of dollars in profits would outweigh a potential environmental problem whose onset and magnitude were unknown.

In their quest to rid the world of these chemicals, commonly used as coolants (such as Freon) for home and automobile air conditioners and in the making of fast-food containers, Rowland and Molina faced a curious dilemma. Since CFCs take 100 years or more to reach the stratosphere, damage to the ozone layer takes place long before it becomes apparent. Prudence suggested that the chemicals be banned before any damage was done. But in the case of the ozone crisis, prudence was forsaken.

Why did it take so long for the world to believe the ozone layer was in danger and act to protect it? The scientists' warning was lost under an avalanche of profits and politics. In 1975, a Du Pont official said, "Should reputable evidence show that some fluorocarbons cause a health hazard through depletion of the ozone layer, we are prepared to stop production of the offending compounds." By 1978, the environmentally conscious spirit of Americans had led to a ban of CFCs in aerosol spray cans—a first step in the phase-out of the dangerous chemicals.

But the vision began to blur. In the early 1980s, a new administration came to power and the leader of the U.S. Environmental Protection Agency, Anne Gorsuch Burford, believed the ozone controversy to be nothing more than a sensational "scare" issue. So many policymakers considered the ozone so unimportant that a Washington environmentalist had to beg his friends to show up at an EPA meeting on the subject. Rebuffed by his colleagues, Sherry Rowland no longer received invitations to speak before industry groups or at other universities' chemistry departments.

Then, suddenly, a calamitous development. Nearly half of the ozone over Antarctica suddenly disappeared in 1985.

It was not until 1988, however, when scientists confirmed that up to 3 percent of the ozone layer over the more populated Northern Hemisphere had been destroyed, that CFC manufacturers finally accepted the evidence. But the damage has been done. As Senator Dale Bumpers remarked in the spring of 1988, "We are now where we should have been a decade ago. Had we acted then rather than reacting now, we would be completing the transition to a new technology today rather than just beginning it."

Although the world has now come to believe Rowland and Molina, many problems and issues remain.

Earlier this year, a group of world leaders proposed a fund to assist developing nations in a CFC phase-out. Officials have long recognized that a complete ban on CFCs would be unachievable for developing nations who wish to modernize and who lack the money and technology to manufacture their own substitute chemicals. The proposal was not supported by the Bush Administration, which balked at the $15 million it would have cost the United States. Each day of indecision, delay, and disagreement increases the price humans will pay for the destruction of the ozone layer. The cost in human suffering and environmental calamity will make $15 million pale by comparison.

In the meantime, for every 1 percent decrease in ozone, non-melanoma skin cancers are expected to rise 5 to 6 percent due to the increase of ultraviolet light. Cases of cataracts and certain human immune-system diseases are also expected to rise. And CFCs are part of an even larger threat. The chemicals contribute to a warming of the atmosphere that could lead to eventual environmental chaos—the greenhouse effect.

Students, policymakers, and anyone who is concerned about the kind of earth they leave to their children and grandchildren should know the story of the 15-year ozone crisis, for it is a lesson for the future. The greenhouse problem, in many ways, mirrors the ozone crisis. The warming of the earth due to manmade pollutants won't be observed until long after the damage has been done. Some scientists believe global warming is already evident. Statistics show that the last few years have been the warmest on record. Could the drought of 1988 be a forecast for our future?

It took a crisis before an international agreement on ozone protection was reached. Will it take another crisis for world leaders to address the greenhouse effect? To avoid the drought, flooding, and disease that will accompany the greehouse effect, world leaders must act now and in unison. Ozone depletion and the greenhouse effect are global problems that do not respect political boundaries.

The human race has already paid a heavy price for its slow response to environmental threats. In hindsight, it is lunacy. Yet our vision of the future is blurry. Said Dr. Wallace S. Broecker, a leader in the study of the greenhouse effect, "The inhabitants of planet Earth are quietly conducting a gigantic environmental experiment. So vast and so sweeping will be the consequences that, were it brought before any responsible council for approval, it would be firmly rejected."

Sherry Rowland and Mario Molina know that the ozone experiment could have been avoided. Instead, we will live with its consequences for another 150 years.

Sharon L. Roan
Orange, California
May 1990

Contents

Timeline

December 1973: Rowland and Molina discover that CFCs can destroy the ozone in the stratosphere.
□

June 1974: Rowland and Molina's paper on their discovery is published in *Nature*.
□

September 1974: Rowland and Molina discuss their theory publicly for the first time at the American Chemical Society meeting in Atlantic City.
□

October 1974: A government committee recommends that the National Academy of Sciences conduct a study on the validity of the CFC-ozone theory.
□

December 1974: First government hearings are held on the CFC-ozone theory.
□

May 1975: The CFC-ozone theory is hotly debated at the American Chemical Society meeting in Philadelphia.
□

June 1975: The Natural Resources Defense Council sues the Consumer Product Safety Commission for a ban on CFCs used in aerosol spray cans.
□

June 1975: Johnson Wax, the nation's fifth largest manufacturer of aerosol sprays, announces it will stop using CFCs in its products.
□

June 1975: A government task force called IMOS defers the decision to regulate CFCs to the pending NAS report.
□

June 1975: Oregon becomes the first state to ban CFCs in aerosol sprays.
□

July 1975: The Consumer Product Safety Commission rejects the NRDC's lawsuit saying there is insufficient evidence that CFCs harm the ozone layer.
□

September 1976: The National Academy of Sciences releases its report verifying the Rowland-Molina hypothesis, but says government action on CFC regulations should be postponed.
□

October 1976: The Food and Drug Administration and Environmental Protection Agency propose a phase-out of CFCs used in aerosols.
□

March 1977: The United Nations Environmental Programme holds the first international meeting to discuss ozone depletion.

□

May 1977: Several government agencies announce joint plans to limit the uses of CFCs in aerosols.

□

July 1977: Harvard scientist Jim Anderson finds an abnormally high level of chlorine oxide in the atmosphere, throwing the CFC-ozone theory into question.

□

February 1978: Government decides to postpone Phase Two regulations on CFCs used in refrigeration, air conditioning, solvents, and other industrial processes.

□

June 1978: The original deadline for Phase Two regulations passes with no action taken.

□

October 1978: CFCs used in aerosols are banned in the United States.

□

November 1979: A second NAS report on the CFC-ozone theory is released, putting depletion estimates at 16.5 percent and saying a "wait-and-see" approach to regulations is not practical.

□

April 1980: The EPA announces the United States' intention to freeze all CFC production at 1979 levels.

□

October 1980: The EPA, under the Carter administration, releases an Advanced Notice of Proposed Rulemaking outlining plans for additional CFC regulations.

□

May 1981: EPA director nominee Anne M. Burford testifies at her confirmation hearings that she views the CFC-ozone theory as highly controversial.

□

July 1981: Hearings are held in Washington to discuss protection of small businesses from possible new CFC regulations. Hearings are highly critical of the Advanced Notice of Proposed Rulemaking.

□

August 1981: NASA scientist Donald Heath announces that satellite records show ozone has declined 1 percent.

□

March 1982: The NAS releases a third report on CFC-ozone and predicts eventual ozone depletion of 5 to 9 percent.

□

March 1983: Burford resigns from EPA. Plans for additional CFC regulations are renewed under new EPA chief William Ruckelshaus.

□

April 1983: During international talks, Norway, Sweden, and Finland submit a world plan for a worldwide ban of CFCs in aerosols and limitations on all uses of CFCs.

☐

February 1984: A fourth NAS report downplays the potential harm to the ozone layer from CFCs by lowering depletion estimates to 2 to 4 percent.

☐

June 1984: At a scientific meeting in Germany, Rowland reports his calculations on heterogeneous reactions involving hydrogen chloride and chlorine nitrate—reactions that could significantly speed up ozone depletion.

☐

November 1984: The NRDC sues the EPA for failing to provide Phase Two regulations on CFCs as specified by the Clean Air Act.

☐

October 1984: A British research group led by Joe Farman detects a 40 percent ozone loss over Antarctica during the austral spring.

☐

January 1985: Lee Thomas takes over as EPA director.

☐

March 1985: The Vienna Convention, calling for additional research and the exchange of information on ozone depletion, is signed by international negotiators. Negotiators fail to agree on worldwide CFC regulations.

☐

May 1985: Farman's paper is published in *Nature*.

☐

August 1985: NASA's Heath shows satellite photos confirming the existence of an ozone hole over Antarctica.

☐

January 1986: A NASA-UNEP report warns that damage to the atmosphere is apparent.

☐

January 1986: EPA releases its Stratospheric Ozone Protection Plan which calls for new studies to determine whether additional CFC regulations are needed.

☐

March 1986: Atmospheric scientists meeting in Boulder discuss plans for an expedition to Antarctica to study ozone depletion there.

☐

June 1986: Papers are published by two research groups indicating chemicals and polar stratospheric clouds are responsible for ozone loss over Antarctica.

☐

June 1986: Hearings on ozone depletion and greenhouse warming are held in Washington. Thomas announces that some government intervention may be needed to halt emissions of gases that could harm the atmosphere. Scientists testify that greenhouse warming has begun due to emissions of gases such as CFCs.

☐

June 1986: CFC manufacturers suggest that safe substitutes for the chemicals might be possible for a high enough price.

☐

August 1986: Thirteen U.S. scientists depart for Antarctica on the National Ozone Expedition.

☐

September 1986: A major CFC industry lobbying group announces it will support limits on CFC growth.

☐

September 1986: The Du Pont Corporation announces it will call for limits on world-wide CFC production.

☐

October 1986: During a press conference from Antarctic, U.S. scientists say they suspect chemicals are to blame for ozone losses there.

☐

November 1986: Scientists favoring weather processes—or dynamical—explanations of the Antarctic ozone loss air their views in a special edition of *Geophysical Research Letters*.

☐

December 1986: International negotiations on ozone protection resume in Geneva after a 17-month layoff. The United States proposes worldwide CFC reductions of 95 percent by the next decade.

☐

March 1987: New evidence supporting a chemical explanation for the ozone depletion in Antarctica is revealed at a scientific meeting in Boulder.

☐

April 1987: Under pressure from some high-level officials, the U.S. backs off on its original position and proposes long-term CFC reductions of 50 percent.

☐

May 1987: Harvard scientist Jim Anderson completes a key instrument for confirming the chemical theories on ozone depletion in time for upcoming expedition to Antarctica.

☐

June 1987: NASA's Heath reports satellite findings of a 4 percent ozone loss detected over a seven-year period. A NASA-sponsored study called the Ozone Trends Panel is organized to review the findings.

☐

July 1987: The State Department announces a Personal Protection Plan as an alternative to CFC reductions. The plan is widely ridiculed.

☐

August 1987: The McDonald Corporation, which uses CFCs in the making of poly-urethane foam containers for hamburgers, announces it will stop using the chemicals.

☐

September 1987: The Montreal Protocol is signed, calling for eventual worldwide CFC reductions of 50 percent.

☐

October 1987: The Antarctic ozone expedition ends with chlorine chemicals found to be the primary cause of ozone depletion.
☐

November 1987: A scientific conference confirms the findings of the Antarctic expedition.
☐

November 1987: U.S. lawmakers call for new negotiations to strengthen the Montreal Protocol.
☐

February 1988: Three U.S. senators ask Du Pont to stop making CFCs.
☐

March 1988: The chairman of Du Pont denies the request to stop making CFCs.
☐

March 1988: The U.S. ratifies the Montreal Protocol in a unanimous vote.
☐

March 1988: The Ozone Trends Panel announces it has found ozone losses of 1.7 to 3 percent over the Northern Hemisphere.
☐

March 1988: Three weeks after refusing to stop making CFCs, the Du Pont Corporation announces it will cease manufacture of the chemicals as substitutes become available.
☐

April 1988: Manufacturers of plastic foam food containers announce they will stop using CFCs.
☐

May 1988: Preliminary findings of a hole in the ozone layer over the Arctic are discussed at a scientific conference in Colorado.
☐

June 1988: A leading scientist says the greenhouse effect is impacting the earth and blames the use of manmade pollutants for the global warming.
☐

August 1988: The EPA orders domestic CFC reductions that mirror the terms of the Montreal Protocol.
☐

September 1988: The EPA says new evidence shows that it underestimated the degree of ozone depletion and says 85 percent cutbacks on CFCs are needed.
☐

October 1988: Scientists meeting in the Netherlands confirm the Ozone Trends Panel findings of ozone losses in the Northern Hemisphere.
☐

March 1989: European countries and the United States agree to faster CFC reductions but developing countries oppose the new timetable citing the costs of substitutes and scientific uncertainty.

To Patrick and Loretta Roan

1
Discovery
January 1973—December 1973

It was a few days before Christmas, and Mario Molina knew he would have a tough time locating his postdoctoral adviser, Sherry Rowland, if he didn't move fast.

Searching the dull green halls of the five-story Physical Sciences Building, Molina's mind raced. It was crazy, the young chemist told himself. The calculations didn't make sense. But he had checked them repeatedly, he assured himself. He had checked them a dozen times. What would Rowland say?

It was 1973 and Molina had only been at the University of California, Irvine, a small but rapidly growing campus 40 miles south of Los Angeles, for about five months. He had completed his Ph.D. for work on chemical lasers in the prestigious chemistry department at the University of California, Berkeley, and was proud of the accomplishment. Chemical lasers was an important field. But now, here he was, only months into his first salaried, faculty-research position, and he had volunteered to study a problem in atmospheric chemistry—something he knew practically nothing about. What if he were wrong about this? Molina spotted Rowland up ahead in the hall and picked up his pace. Then again, he drew in a deep breath, what if he were right?

Rowland looked harried. He was about to depart for a five-month sabbatical in Vienna. The holidays were nearing and he had much to do before he left. But Rowland was known for his patience and his ability to get things done under extreme pressure. He gave Molina his full attention.

"Hey, we better talk before you leave," Molina said. "I may have found something important."

The two conferred for a few minutes and agreed to meet again later that afternoon. Within hours Rowland had cleared his busy schedule and settled into his cluttered corner office to go over Molina's calculations. While showing no outward emotion in front of Molina, Rowland was stunned. He, too, thought there must be an error. Where was it? Why couldn't the two of them find it? This was just too implausible.

The two men poured over the calculations for the next two days, not once discussing the implications of the work. And yet both of them knew that the equations they had scribbled could reveal a profound, almost unthinkably horrible secret about the earth's delicate atmosphere. If their work was correct, then a common class of manmade industrial chemicals called chlorofluorocarbons, or CFCs, signaled big trouble for the environment and, inevitably, the human race.

CFCs were widely used as aerosol spray propellants and refrigerants and, in 1973, were being produced worldwide at the rate of almost a million tons per year. People assumed that CFC emissions were harmless. But Rowland and Molina's calculations showed that the gases would eventually rise to the upper atmosphere and begin destroying ozone, the thin layer of gas that rests about 8 to 30 miles above the earth and shields it from the searing effects of the sun's ultraviolet light.

Imagine an eighth of an inch of snow dusting the entire world. The ozone layer is only as thick as that. This trace of ozone protects the earth. And yet its role is essential to the continuance of life. Now, Rowland and Molina suddenly realized, human activities could be destroying it and doing the job rather quickly. Rowland left his office that night and went home without the feeling of excitement that one would expect from a scientist who had toiled for years doing basic research and had now made a major discovery. Instead, he felt something of a burden.

"The work is going well," he told his wife, Joan. "But it looks like the end of the world."

□   □   □

"I got a chilly feeling," Rowland told a *Newsday* reporter shortly after the discovery became public. "Suddenly, something runs down your spine. It's no longer the same thing; it's not the same circumstances as the usual experiment. You see potential danger for the environment of the world. My immediate reaction was disbelief. I thought I must have done something wrong."

A chemical reaction that could trigger a dramatic shift in the earth's superbly balanced climate was not what Rowland or Molina expected to find when they began looking at the chemicals in the fall of 1973.

The idea of studying CFCs had been nothing more than an interesting idea when it occurred to Rowland during a meeting sponsored by the Atomic Energy Commission in balmy Fort Lauderdale, Florida, almost two years earlier. As a radiochemist who had devoted his career to the study of how small molecules react and decay, Rowland had received funding for his research from the Atomic Energy Commission for years. The January 1972 meeting had been organized in order to bring together chemists and meteorologists—a gathering that seemed as odd as a bunch of pediatricians and gerontologists meeting to chat about the latest trends in medicine.

"There's a great difference in background," said Rowland of the two disciplines. "The typical chemist has had no courses in meteorology and the typical meteorologist has had maybe one course in chemistry."

It was not until the mid-1980s that one would expect a meteorologist to have had some training in chemistry. And the fact that, in 1972, meteorology and chemistry had not yet begun to merge may have been one reason no one had thought to look at what happened to manmade chemicals released in the upper atmosphere; the stratosphere was sort of a no man's land among atmospheric scientists. The thought certainly hadn't occurred to Rowland as yet, either, as he departed for Fort Lauderdale. Rowland had no background in atmospheric chemistry but was intrigued with the idea of applying his knowledge in radiochemistry to another field. He decided to attend the meeting to scout for some new ideas.

As was usual during scientific conferences, the coffee room chatter proved to be more interesting than some of the formal presentations. During one break between the talks and after a presentation on CFCs, Rowland heard an intriguing bit of news regarding work by the British scientist James Lovelock, a fiercely independent and creative chemist who had invented a much-heralded device, called an electron capture gas chromatograph, to detect atmospheric gases in minute amounts. Using his invention, Lovelock had, in 1970, discovered a chlorofluorocarbon chemical called trichlorofluoromethane in the air over western Ireland—a surprising discovery since most pollutants rarely lingered for long in the atmosphere. Following up on that finding, Lovelock had discovered two commonly used CFCs—called CFC-11 and CFC-12 according to chemical industry shorthand—in the atmosphere at concentrations of 230 parts per trillion, which was equal to about one drop in a large swimming pool. Lovelock saw no reason to be alarmed by the apparent long, vigorous life of the chemicals in the atmosphere, however, noting that the highly stable CFCs might be useful to show how air masses moved. (Ironically, Lovelock's family was the first victim of the discovery that CFCs persist stubbornly in the air. Long before aerosol sprays containing CFCs were banned in the United States, the Lovelocks were forced to give up using aerosol sprays because they interfered with the measurements Lovelock was trying to make in his laboratory at home.)

The thing that struck Rowland as odd was that the concentrations of the gases Lovelock had measured appeared to be close to the total amount of CFCs being produced, suggesting the chemicals were drifting lazily up through the atmosphere with amazing endurance. The atmosphere, scientists knew, had a beautiful ability to cleanse itself of the earth's byproducts. Most chemicals were eventually washed out of the atmosphere through "sinks," removal processes in the lower atmosphere such as rain. But CFCs were apparently immune to these normal removal processes. Rowland returned to Irvine wondering what eventually became of the chemicals. But the question was not of pressing importance. He simply filed the matter away and didn't resurrect it for more than a year.

For the 46-year-old Rowland, having a stockpile of new problems to mull was crucial to his success as an academic chemist. If there was anything he feared in his career it was getting into a rut. And so far, he had successfully avoided that.

Born in Delaware, Ohio, Rowland was a fast learner who breezed through his studies at Ohio Wesleyan University where he majored in chemistry and minored in journalism. (The minor in journalism would later serve him well in his dealings with the press.) At 6 feet 5 inches, he wasn't too bad at either basketball or baseball and played both in college before moving on to complete his master's and doctoral degrees in chemistry from the University of Chicago. During graduate school, Rowland specialized in radiochemistry and had the good fortune of learning his discipline from Willard F. Libby, the Nobel-prize-winning scientist who invented carbon 14 dating. After four years at

Princeton University, Rowland accepted a faculty position at the University of Kansas and developed a reputation in radiochemistry, traveling throughout the world to attend scientific conferences in his specialty. By the time he agreed to head up the new chemistry department at the University of California, Irvine, in 1964, at the age of 36, he was relatively well known in his field.

"A well-known unknown," he called himself.

But in the summer of 1973, Rowland was anxious to study "something where the questions are a little bit different and the answers are a little more unknown." At UCI, he directed a small but eager group of young chemists. One of them, Mario Molina, showed particular promise and was also looking around for a research project.

The two men had met a few months earlier at a science conference, and Rowland was delighted when the young chemist chose Irvine to undertake his postdoctoral work in photochemistry. Molina was born in Mexico City, the son of an ambassador. Because of his obvious intelligence and his family's connections, Molina's career path was carefully laid out for him. He received some of his early education in Switzerland and earned a bachelor's degree in chemical engineering at the University of Mexico. After graduating, Molina spent almost a year in Germany at the University of Freiburg and later studied at the Sorbonne. The rather informal graduate-school programs in Europe were not to his liking, however, and Molina chose Berkeley to work on his Ph.D. in order to study under a highly respected chemist named George Pimentel. By the time he joined Rowland's research group in October of 1973, Molina was in his late twenties.

It was Rowland's practice to offer his postdoctoral students a choice of problems to study. One of the problems was tracing the whereabouts of CFCs in the atmosphere, and Rowland was pleased when Molina selected the topic and set off happily to try to come up with some answers.

"We thought it would be a nice, interesting, academic exercise," Molina said later. "We both knew that these CFCs were rather stable so there was nothing obvious that would damage them soon after they would be released. But that's about as much as I knew at the time."

Molina, a slightly built man with a calm demeanor and intense brown eyes, proceeded to tackle the problem in his usual studious and methodical manner. For several weeks, he spent much of his time in the campus library, reading books on atmospheric chemistry and proceeding to the more sophisticated scientific literature. Part of his preparation was to look at how other compounds reacted in the atmosphere. Eventually, he narrowed his focus to CFCs.

Molina knew that the total amount of CFCs in the troposphere—the part of the atmosphere between the earth's surface and about 6 to 10 miles in altitude where weather patterns formed—was about equal to the total amount produced. That was what Lovelock had found. So it appeared that the normal removal processes for gases reaching the troposphere didn't affect CFCs. Molina knew, for example, that rain would have to be ruled out as a disposal mechanism because CFCs were relatively insoluble in water. Finding no other removal processes for the chemicals in the troposphere, Rowland and Molina assumed that the chemicals were reaching the stratosphere, the next layer of the atmosphere, above the troposphere.

As far back as the Fort Lauderdale meeting it had occurred to Rowland that if CFCs reached the stratosphere they would be broken up by short-wavelength ultraviolet light. Since the chemicals do not absorb the longer wavelengths of sunlight they would not be attacked in the lower atmosphere. But, once they passed the protective ozone blanket, short-wavelength radiation would easily break them down into other compounds and atoms in a process called photodissociation. In particular, Rowland and Molina deduced that the photodissociation would allow for the release of free chlorine atoms from CFCs. They also estimated that the CFCs' journey to the stratosphere was a slow one. They calculated that it would take between 40 and 150 years for chemicals released on earth to reach their demise.

Satisfied, the two chemists concluded that there was no sink for CFCs other than the stratosphere and thought about writing up the results of the work for publication. But, Rowland told Molina, "We might as well be complete about it and find out what's going to happen to the chlorine."

Molina now had to consider how the chlorine would react with the thin, wafting layer of ozone in the stratosphere. Ozone is the product of sunlight acting on oxygen. While ordinary oxygen molecules have two atoms, ozone is made up of three oxygen atoms and is unstable, meaning that it is always willing to give up one of its oxygen atoms to other gases and turn back into oxygen. It is created when ultraviolet radiation or discharges of electricity split oxygen molecules

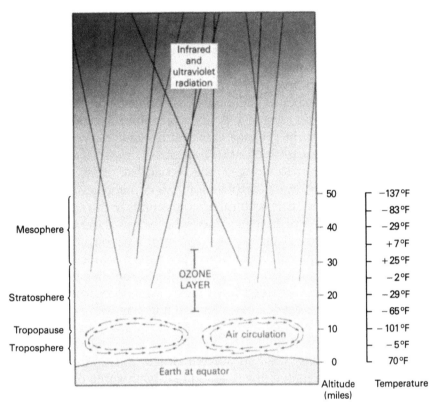

The earth's atmosphere is comprised of a series of layers at which different natural phenomena take place. The troposphere is the layer nearest the earth. The next layer, the stratosphere, is about 13 to 35 miles in altitude. Here, the ozone layer absorbs the sun's ultraviolet light and temperatures begin to rise with altitude. The next layer is the mesosphere where temperatures begin to fall again. The ionosphere marks the upper boundary of the atmosphere.

into single oxygen atoms. The oxygen atoms then link with oxygen molecules to form ozone. Derived from the Greek word *ozein*, meaning "to smell," ozone at ground level gives off a pungent, acrid odor that can sometimes be detected around leaking power lines, electric toy trains, or after a lightning storm.

Although it is no more substantial than a sheet of tissue paper wrapped around the earth, ozone protects life on earth. While ozone found at ground levels contributes to smog, the gauzy layer in the stratosphere has two purposes, and both of them are beneficial to living things. First, ozone absorbs much of the sun's harmful ultraviolet light and is the only substance that does so. Second, by absorbing some of the sun's rays, ozone creates the stratosphere—a layer of the atmosphere in which temperatures rise with altitude thereby regulating worldwide circulation patterns and keeping "weather" confined to the troposphere. Without ozone, temperatures in the atmosphere would gradually decrease. While ozone is constantly produced by the action of sunlight on oxygen it is not a plentiful substance. If ozone were compressed to the pressure at the earth's surface, it would consist of a layer only one-eighth of an inch thick.

As Molina began working on the second half of his assignment, the fact that ozone was unstable raised some interesting questions about how it would react with other compounds in the stratosphere. But Molina could never have imagined what he was about to discover.

Using detailed calculations for chemical reactions, Molina concluded that each chlorine atom from CFCs would collide with a molecule of the highly unstable ozone. But the action didn't end there. Once the chlorine was freed from the chlorofluorocarbon, the by-product would be oxygen and a chemical fragment with an odd number of electrons called chlorine monoxide. Chlorine monoxide is a free radical, an unstable and highly reactive fragment with an odd number of electrons. The odd number of electrons in the fragment, Molina knew, guaranteed that it would react with a free oxygen atom in order to achieve an even number of electrons. Thus, he calculated, when the chlorine monoxide met the free oxygen atom, the oxygen in the chlorine monoxide would be attracted to the free oxygen atom and split off to form a new oxygen molecule. Chlorine would then be freed which would collide with ozone, thus starting the cycle all over again.

In short, the breakdown of CFCs by sunlight would set off a catalytic chain reaction in which one chlorine atom, like a monster Pac Man, could gobble up 100,000 molecules of ozone, turning them into impotent molecules of oxygen. Molina knew there were other catalytic reactions that occurred in the atmosphere—reactions in which one compound, the catalyst, could destroy another compound and would not be destroyed itself. But, if his calculations were correct, it looked as if the chlorine catalyst was an especially vigorous one.

Visions of environmental disaster hadn't yet marred Molina's thoughts, however. So the chemicals could take out a little ozone. What he really needed to know was just how much of the stuff was being dumped into the atmosphere so that he could estimate how much ozone might be lost. He knew other trace gases were released into

To be exposed to ultraviolet light, the CFCs must move upward, beyond the ozone layer. There, the UV light breaks the compounds down releasing chlorine atoms (Cl). The free chlorine atoms then attack ozone as follows:

$$Cl + O_3 = ClO + O_2$$

$$ClO + O = Cl + O_2$$

effect: $O + O_3 = 2O_2$

In the first reaction, chlorine atoms destroy ozone to form chlorine monoxide (ClO) and normal oxygen (O_2).

In the second reaction, the Cl atom is regenerated and is free to start the cycle over again. The atom of oxygen that is lost in the second reaction is the equivalent of another ozone molecule lost because, otherwise, O and O_2 would have formed ozone (O_3).

The final effect is that two molecules of ozone are consumed each time the cycle is repeated. Chlorine is freed in each cycle to begin the process over again.

How chlorine destroys ozone.

the atmosphere all the time through natural mechanisms and that the biosphere continued to maintain its harmonious balance. It was unlikely, he figured, that a manmade chemical could outmuscle natural reactions in the atmosphere.

"I was thinking that (the effect) would be negligible because one normally thinks of natural processes on a global scale as being much larger," Molina said.

But nature was about to be outdone by the chemical industry. Checking industry figures for the amount of CFCs produced in 1972, Molina converted the figures into scientific units. He stared at his note pad and immediately suspected an error.

He rechecked the figures and, finding nothing amiss, began leafing through industry production figures to compare the amount of CFCs produced to another chemical, such as the nitrogen oxides—nitric oxide and nitrogen dioxide—which are produced in the atmosphere from nitrous oxide. Known as laughing gas, nitrous oxide was also used as an aerosol propellant. Molina saw that, compared to the production of nitrous oxide, CFC production was overwhelming.

He had to find Rowland fast.

Within a couple days, Rowland and Molina had reached a conclusion about the effects of CFCs in the atmosphere. They calculated that the amount of CFCs released that year—almost a million tons world-wide—would reach the stratosphere some 100 years later. Most of the CFCs released since the 1930s, when they were first manufactured, hadn't yet reached the mid-stratosphere where the reactions could take place. Significant production of the chemicals hadn't even begun until the 1960s. So actual damage to the ozone layer wasn't yet apparent. The concept was based on a principle called steady state—the condition of the atmosphere in about 100 years, when the amount of CFCs destroyed in the atmosphere by sunlight would be equal to the rate at which they were released into the atmosphere. In other words, the amount of ozone being destroyed would be equal to the amount being created.

The chemists determined that, at 1973 levels, between 7 and 13 percent of the ozone layer would be depleted within about 100 years— enough ozone depletion to seriously alter life on earth.

It was just a few days before Christmas and Rowland now faced a difficult decision. He and his wife Joan were scheduled to depart for Vienna just after New Year's, thanks to a Guggenheim Fellowship Rowland had earned. It seemed the perfect irony, that, with a problem of potential global importance sitting before him in Irvine, he was about to go off on a sabbatical to look for something novel to do. He had all he could cope with in the Irvine chemistry department.

Before making any hasty decisions, however, Rowland decided to share the CFC discovery with a few people whose opinions he trusted. One was Dieter Ehhalt, an atmospheric chemist from Germany who was working at the National Center for Atmospheric Research in Boulder, Colorado, at the time. Ehhalt had done some similar work collecting air samples from balloons in order to look at trace gases and had attended the meeting in Fort Lauderdale where CFCs were discussed. Rowland telephoned Ehhalt and told him that he was interested in getting some of his air samples. "What for?" Ehhalt inquired.

"We've been doing some calculations on chlorofluorocarbons and it looks very interesting. There are very large concentrations of it," Rowland told him.

Ehhalt was puzzled. "There isn't very much of it there now."
"We've gone to steady state," Rowland answered.
Ehhalt paused. "Beautiful," he answered.
Immediately, Ehhalt had understood the implications of Rowland and Molina's calculations.

The other phone call Rowland made that week was to Al Wolf, a radiochemist at the Brookhaven National Laboratory who later developed the radioactive chemicals that serve as the basis for PET scanners used to diagnose illnesses. Rowland, who had known him since the early 1950s, called Wolf, explained the research, and told Wolf that it looked like there was a major threat from chlorofluorocarbons. Wolf's reaction was probably the soundest piece of advice Rowland was going to hear for a long, long while.

"Well," said Wolf, matter of factly, "I can tell you Du Pont's not going to stop making them."

□   □   □

There was one more person that Rowland and Molina wanted to see before making their findings known to the world. During the week between Christmas and New Year's, the pair boarded a plane for San Francisco to visit Harold Johnston, a chemist at the University of California, Berkeley.

Johnston was a courteous, Southern gentleman who could be very tough and outspoken. He was, perhaps, the perfect person for Rowland to speak to about his CFC-ozone theory. An expert in atmospheric chemistry, Johnston had studied the growing smog problem in Los Angeles in the 1950s. He was the author of several books on the atmosphere and was extremely knowledgeable about ozone and chemical reactions in the stratosphere. He also knew a thing or two about the kind of "reactions" that could be set off by mixing industry matters with the environment and politics. In 1971, Johnston had suggested that gases called nitrogen oxides, released in the exhaust from high-flying supersonic jets, could cause ozone depletion through the same kind of catalytic reaction Rowland and Molina were now talking about—a theory that had generated intense disagreement and, in 1973, was still the prime topic of discussion in atmospheric chemistry. Due to his research, Johnston had suffered occasional personal attacks from SST proponents.

Because he wasn't an atmospheric chemist, Rowland was only

slightly familiar with the debates over the fate of the proposed fleet of the tilted, dropped-nose aircraft. Supersonic jets, or SSTs, could bridge any two points on the earth within 12 hours, and the idea that environmental consequences might squelch this wondrous new technological development led to the acrimonious debate. Rowland had invited Johnston to speak on the Irvine campus a couple of times and knew that the ozone-SST issue was controversial. He knew Johnston could bring him up to date with what was known about chemical reactions involving ozone.

Before the debate over nitric oxides from SSTs, no one had really given much thought to the ozone layer and its sustenance. In 1930, an English scientist named Sydney Chapman attempted to explain how ozone was formed and destroyed in the atmosphere. Called the "Chapman mechanism," the scientist suggested that ordinary oxygen molecules (O_2) would absorb very short-wavelength ultraviolet light. Oxygen molecules make up 21 percent by volume of all atmospheric molecules while nitrogen takes up the bulk at 78 percent. Sunlight, Chapman said, would split apart oxygen molecules into two oxygen atoms (O). Atoms of oxygen would then attach themselves to other oxygen molecules to form ozone (O_3). Chapman also proposed that oxygen atoms could break up ozone molecules by colliding with them to produce two oxygen molecules. So, the Oxford scientist concluded, the rate of ozone being produced was equal to the amount being destroyed at any one time.

Until the SST debates and the studies they generated, few people were interested in processes that took place in the mysterious stratosphere. Meteorologists, for example, were usually more concerned with weather, which took place in the troposphere. People who launched rockets were more concerned with processes in the mesosphere—the layer of the atmosphere above the stratosphere. But, from the 1950s to the 1970s, a small enclave of scientists who met occasionally to discuss ozone studies concluded that the Chapman mechanism was not complete.

In the late 1950s, rocket measurements showed that the amount of ozone present in the stratosphere was much less than what theory suggested. By 1961, ozone researchers suspected that something else must be destroying ozone. In 1964, a Canadian researcher named John Hampson proposed that oxides of hydrogen, which were free radicals given off from water vapor, could be created in the stratosphere and could react with ozone in a catalytic reaction; the ozone was destroyed in the reaction but the catalyst, the oxides of hydrogen, was not. A few years later, Australian scientist B.G. Hunt set up a computer model showing how oxides of hydrogen might account for the apparent ozone loss.

In 1969, however, the brilliant meteorologist of the University of Stockholm, Paul Crutzen, began to study the idea that activities on earth could affect the atmosphere. While on a sabbatical at Oxford, Crutzen, who had studied chemistry, wrote a paper that received very little attention suggesting that the oxides of hydrogen reaction was not important. In a second paper, a year later, Crutzen suggested that nitrous oxides, which could be emitted from natural sources such as solar flares, might be responsible for some ozone loss. Crutzen suggested that nitrous oxides, because of their inertness, would rise slowly to the stratosphere where ultraviolet radiation would convert the nitrous oxide to nitric oxide, a simpler form of the chemical. The nitric oxide would then initiate a catalytic reaction that could destroy ozone while leaving the nitric oxide molecules free to begin the chain reaction over again.

Crutzen's theory was interesting to some people but didn't generate much excitement. The source of nitrogen was small, too small to do any great harm to the ozone layer. But the work was to become more than just a nice bit of academic research. In 1971, congressional committees began to hear scientific arguments on the potential problems of SSTs (which released oxides of nitrogen) and to consider a bill to provide funding for the construction of two supersonic jet prototypes by the Boeing Company.

The SSTs were generating a lot of controversy. A study authorized by the Department of Commerce had found the jets posed no threat to ozone. But critics charged that the aircraft would be noisy and costly. And there was another potential problem. An atmospheric scientist at the University of Arizona, James McDonald, was suggesting that water vapor from the jets' exhaust could harm the ozone layer, allowing more ultraviolet light to reach the earth and causing an increase in skin cancers. Besides being one of the first scientists to suspect that manmade pollutants could disrupt the ozone layer, McDonald was perceptive enough to link the increased ultraviolet light from ozone depletion to an increase in skin cancers. However, the outspoken scientist suffered for his independent thinking. The link between skin cancer and the sun's radiation was not widely known in the early 1970s. And the suggestion didn't impress lawmakers when McDonald testified at congressional hearings on the SSTs in March of 1971.

Unfortunately, McDonald was an easy target for ridicule. A believer in UFOs, he had once suggested that alien visitors might be responsible for power blackouts on earth. But, in the spring of 1971, he only wanted to warn Congress of his theory that water vapor from the exhaust of SSTs could destroy the ozone layer. Committee members grilled McDonald mercilessly, however. In their book *The Ozone War*, authors Lydia Dotto, a Canadian science writer, and Harold Schiff, a chemist, said committee members questioning McDonald seemed more concerned with his views on UFOs.

"It is not entirely clear that there is a relationship between SSTs and UFOs," McDonald protested to the committee, straining to be patient.

McDonald's theory was sound enough to cause considerable alarm among environmental groups and government officials, however. In March 1971, the Department of Commerce organized a scientific conference to study the impact of the SSTs on the atmosphere. Hal Johnston was invited to attend the meeting.

Johnston knew little about the SSTs at that point and listened with only mild interest as the theory of water vapor was discussed during the two-day meeting. McDonald's water vapor theory was soundly refuted during the meeting and was later completely disproven. But when someone raised the issue of whether oxides of nitrogen, also released in the jets' exhaust, might have some potential effect on ozone, Johnston perked up. He had earned his Ph.D. on a study of oxides of nitrogen and was familiar with the compound. Conference participants quickly dismissed the idea that oxides of nitrogen could be important in the atmosphere but Johnston knew better. He spoke up quickly, disagreeing with the consensus. But, he recalled later, "Nobody wanted to hear that at that meeting. There was quite an uproar over it."

Johnston returned to Berkeley incensed that the oxides of nitrogen mechanism was being ignored and went to work on a paper describing how the compound, released in the exhaust of supersonic jets, could cause a catalytic reaction in the stratosphere which would destroy ozone. He deduced that 500 SSTs flying seven hours a day would deplete ozone by at least 10 percent in one year's time. The paper was accepted by the respected journal *Science* and was published in August 1971. But by then it was old news. In a scandalous mixture of science and politics, someone (the culprit was never found) leaked a version of Johnston's paper to a small Bay Area newspaper in May, only days before a crucial Senate vote on whether to fund the

SSTs. With Johnston's findings out in the open, it was not surprising that the Senate defeated the SST measure.

Johnston's paper in *Science* was followed by a paper from Crutzen, also suggesting that oxides of nitrogen could have a major impact on the atmosphere. Concerned with the recent flurry of discoveries on the environmental impact of high-tech aircraft, the federal government ordered a multimillion-dollar study from the Department of Transportation to assess the threat of supersonic aircraft being developed by other nations. The Climatic Impact Assessment Program (CIAP) became the first intensive study of the effects of human activities on the atmosphere.

Ironically, research later showed that nitric oxides from supersonic jets were far less of a threat to ozone than had been previously suggested. The jets would fly too low to cause much damage. (Today, most scientists believe that a large fleet of high-altitude aircraft could cause substantial ozone depletion, however such a fleet has not been built. And there is controversy over the Department of Commerce's current position that the proposed "Orient Express" would not cause losses of ozone.)

Water vapor from the exhaust was never proposed as a possible threat to ozone again. Although no one would ever know if the grilling he took earlier that year contributed to his depression, James McDonald was found dead by suicide in mid-1971.

□ □ □

Studies for the Climatic Impact Assessment Program were well underway by the time Rowland contacted Johnston in December 1973. Johnston, a respected chemist, listened to Rowland's explanation of the CFC problem over the phone and, to Rowland, seemed interested although noncommittal. He would have to see more details of the work and review it. In his office, Johnston hung up the phone while experiencing the same conflicting emotions that Rowland and Molina had felt a few days earlier.

"That's terrific. That's terrible," Johnston muttered, impressed with the theory and yet horrified by its implications. But, he said later, "You could see immediately that it was very important."

Within days the Irvine scientists met in Johnston's Berkeley office and waited while Johnston reviewed the data carefully. He could find no errors in the work. In fact, Johnston informed Rowland and Molina that they weren't the first to look at the reaction of free chlorine in the stratosphere.

Only a few months earlier, while Rowland and Molina were just beginning their work on CFCs, a CIAP meeting had been held in Kyoto, Japan. Attending the meeting was a young University of Michigan researcher named Richard Stolarski who was presenting a study completed by he and another Michigan scientist, Ralph Cicerone. Neither was an atmospheric chemist. Stolarski was a physicist and Cicerone had trained in electrical engineering with a minor in physics, but they were eager to study stratospheric problems and had begun a study under contract for NASA's Marshall Space Flight Center to determine the environmental effects of the proposed space shuttle. In particular, the pair decided to look at the hydrogen chloride emissions from the solid-fuel rockets.

Stolarski and Cicerone were young, inquisitive, and very sports-minded, often lugging their golf clubs to science conferences in the hopes of squeezing in a round after the meetings. Stolarski and Cicerone had been attending the CIAP meetings, had heard about the SST controversy with nitric oxide, listened to the wrangling over stratospheric chemistry, and thought the whole field was rather interesting. They wanted to get involved. And no one had thought about chlorine chemistry before.

"Chlorine chemistry was a neat way to get involved in it because it exhibited all the same kinds of things that the nitrogen oxide chemistry did but nobody cared about it," Stolarski later recalled. "So we thought we could very quietly work on it and not get involved in the incredible brouhaha that was going on then about SSTs. And if you were new in the field and weren't quite as sure of yourself, you didn't like to expose yourself until you learned a little better. It was a nice, quiet way to learn. But it didn't stay quiet very long. It just blew up in our faces. But it got us in the field."

In 1973, Stolarski and Cicerone had determined that the chlorine released from exhaust from the shuttle might deplete ozone. Initially, NASA officials seemed unconcerned with the findings. "Of course, chlorine will catalyze ozone destruction," the researchers were told by a NASA official. They were reminded that in the 1930s researchers had combined chlorine and ozone in the laboratory and had subjected it to light and had observed the ozone decrease. It was a known fact. In addition, the Michigan scientists' calculations showed the amount of ozone depletion that could be expected from the shuttle exhaust was a tenth of a percent.

It didn't seem to be a major problem. But NASA officials had to think about chlorine in combination with other environmental effects from the shuttle. And when Stolarski and Cicerone informed NASA of their findings and discussed publishing the work, some NASA officials said they preferred to keep the shuttle research quiet.

"There were all kinds of opinions within NASA as to what stand to take on something like this," Stolarski recalled. "There were certainly people early on trying to convince us that perhaps we didn't want to publish some of the things we found."

NASA, however, remained supportive of the research, Stolarski remembered. "Although the guys that supported us were, at times, a bit reluctant to advertise it."

Trying to be complete about their studies on chlorine in the stratosphere—and perhaps confused that they had touched a political nerve—Stolarski decided not to mention the effects of ozone depletion from the shuttle at the CIAP meeting in Kyoto. Instead, he suggested that chlorine, produced from natural sources such as volcanoes, could create a catalytic reaction to destroy ozone. The finding was of some interest, but not because anyone was worried about a sudden rash of volcanic eruptions.

Cicerone and Stolarski later spelled out the full consequences of chlorine from the space shuttle exhaust in a paper they submitted to *Science* magazine. After a long wait, the *Science* paper was rejected, having received a rather mixed and mediocre review. Cicerone immediately suspected some competitors had done the paper in, aghast over one comment made in the review of the paper which said, "This work is of no possible geophysical significance."

A special edition of the *Canadian Journal of Chemistry* was about to be published summarizing the proceedings at the Kyoto meeting, however. And although Stolarski and Cicerone were depressed about the rejection of their paper by *Science*, they were persuaded to submit it to the Canadian journal. The study appeared along with one by two Harvard researchers, Michael McElroy and Steven Wofsy, who had reached similar conclusions about chlorine from the shuttle after hearing Stolarski discuss the problem in Kyoto. Along with Rowland and Molina, Stolarski, Cicerone, McElroy, and Wofsy were all to become key participants in ozone-depletion arguments throughout the 1970s and 1980s.

The secret about the effects of the space shuttle's exhaust was out. But, Cicerone and Stolarski concluded, the spaceship was still on the drawing board and probably wouldn't release enough hydrogen chloride to do much harm. Johnston suggested another version of NASA's reaction to the chlorine studies: "The people running the space shuttle program were trembling in their boots for a year or two to keep that from coming under the same kind of politics the SST ran in to," he said with characteristic candor.

The chlorine issue was dropped. Other than volcanic emissions and exhaust produced by the space shuttle, which wasn't even built yet, there was no other known source of free chlorine in the stratosphere.

But, six months later, Stolarski, a self-effacing person who is now known as a top researcher in the field, was kicking himself for limiting his thoughts to those two known sources of chlorine. There had been several hints about additional chlorine sources in the early 1970s that the two Michigan researchers had ignored.

In November 1973, Stolarski had attended a meeting in Houston where he encountered Charles Kolb, a scientist from Aerodyne in Cambridge, Massachusetts. Kolb had heard about the chlorine discussions at the Kyoto meeting and now pulled Stolarski aside in Houston.

"Have you thought about the Freons?" he asked Stolarski. "They are incredibly stable molecules and they've got to be a good source." Stolarski, who was not a chemist, had no idea what a Freon was.

"Well, no, I hadn't really thought about it," he stammered, shrugging off the suggestion.

There was a second hint that also fell by the wayside. Stolarski had been discussing the chlorine research on occasion with Jim Walker, a Yale scientist. In November, Walker sent Stolarski a one-page Xerox of Lovelock's carbon tetrachloride measurements.

"Have you thought about carbon tetrachloride?" Walker had scribbled across the top. Stolarski and Cicerone hadn't and let the matter drop.

There was just too much work to do to look into every little suggestion. But Stolarski had no doubt that the chlorine research was important.

"Why are you studying the chlorine from the space shuttle?" a colleague asked Stolarski that winter. "It's such a trivial effect. You're calculating a tenth of a percent. Who cares?"

Stolarski's response was one of the few times, he recalled, that he has been on the money.

"We study it because it's interesting," he answered. "And some day, somebody is going to invent a much bigger source that's going to be important."

Of course, somebody had long ago invented that much bigger source. And, unknown to Cicerone, Stolarski, Wofsy, McElroy, and others, two somebodies had just found it.

□   □   □

Johnston peered at Rowland and Molina in his Berkeley office and informed them that they had just found the first major source of free chlorine that warranted serious concern.

Having participated in the SST-ozone controversy, he knew the Rowland-Molina theory would stir up as much debate and inevitable rancor as the SST thing. He advised Rowland and Molina to begin making their case. The Irvine scientists obviously felt a deep moral conviction that they should speak out firmly and immediately about the potential dangers of CFCs. Because Johnston was familiar with such matters, Rowland and Molina inquired whether he would care to join them in raising the red flag on CFCs.

"No," Johnston told them, with a grin. "That's up to you." But before Rowland and Molina caught an evening flight back to Orange County, Johnston left them with some words to mull over.

"Are you ready for the heat?" he asked.

2
Death to Ozone
January 1974—December 1974

The way things stood now, Rowland wondered if he would even get the chance to feel any heat if he were to go on to Vienna for the sabbatical.

After discussing the matter with Joan, the couple decided to proceed with their travel plans. Rowland would discuss the CFC-ozone theory in Europe while Molina continued working on the theory at Irvine.

Arriving in Vienna, Rowland quickly located an apartment and spent his third day in the city writing a paper on the theory. He submitted the paper to the prestigious British science journal *Nature* and began a five-month wait—a period he likened to waiting for a very slow fuse to burn down on a bomb.

Rowland was eager to discuss the work, however, and decided to talk about the findings in a series of lectures he was scheduled to give while in Europe. At one meeting before a group of meteorologists in Sweden, Rowland received a taste of how he would be required to defend the ozone-CFC theory over and over again in the coming years. The day before he was to give his talk, Rowland had met Paul Crutzen for the first time and discussed the theory with him. Crutzen had read the paper Rowland was about to discuss and raised a question about the calculations. Concerned that a flaw in the theory was being unveiled only hours before he was to be grilled by his audience of meteorologists, Rowland sat up most of the night in his hotel room punching numbers into his calculator. In his apartment, Crutzen was doing the same. Meeting the next morning, however, the two had reached the same conclusion; Crutzen had found a minor point to the theory that had needed clarification but it hadn't changed the essence of Rowland and Molina's findings. It still looked as if deodorants and hair sprays could wreak havoc in the atmosphere.

Rowland was to receive another curious challenge to the theory while in Europe, although it was hardly among the most threatening he would encounter. Before the paper appeared in *Nature* in June, a Swedish newspaper ran a story on the theory, having discovered it from preprints of the article that Rowland had sent out to Crutzen in February. (Preprints are copies of a scientific paper made available after the paper has been submitted to a journal but before it is published.) Crutzen had mentioned the paper in a talk in Stockholm with a newspaper reporter who quickly pounced on the ozone-CFC connection. In the original paper, Rowland and Molina had referred to the CFCs by their most common trademark name, Freon, a trademark that belonged to the Du Pont company, the world's largest CFC manufacturer. And, shortly after the Swedish newspaper story appeared, Rowland received a telephone call from a Du Pont public relations executive in Europe. The executive was appalled at what he viewed as a ludicrous theory. He also demanded to know what Rowland was thinking of by using the trademark name Freon.

"His main concern seemed to be that I was using a trademark name. I took that very seriously and switched to 'chlorofluorocarbons' from then on," Rowland later recalled. The man said nothing about Rowland and Molina's theory that his company might be contributing to the downfall of modern man.

Rowland heard from the Du Pont executive one more time in the summer of 1974 after returning from Vienna. Again, environmental disaster and who might be responsible for it was not the issue. The Rowland-Molina paper had appeared in *Nature* on June 28, 1974. Why then, the Du Pont official demanded to know, did a Tehran newspaper publish a story on the CFC issue the first week of June? Although it happens frequently, and sometimes inadvertently, scientists prefer that their work not be published in the popular press before it has appeared in professional journals. The Du Pont executive demanded to know what Rowland thought of the ethics of revealing a story three weeks before *Nature* published it.

□   □   □

Rowland couldn't quite understand how the Tehran newspaper had picked up the story before it was published in *Nature*. Contacting a friend in Tehran, Rowland had him check the date of the story and found that the story was a Reuters news service piece that had appeared the first week of July, after the *Nature* paper. The Du Pont executive had the dates wrong. Well, Rowland concluded, the poor man must be very confused indeed if all he could think of in the aftermath of the *Nature* paper was when the readers of Tehran became informed of the problem.

Sitting down to write a letter to the Du Pont official informing him of the actual date of the Tehran story, Rowland could only wonder how someone could so completely miss the point. "It didn't strike me that this was an ethical concern that was anything like the importance of putting molecules in the atmosphere that were going to destroy the ozone layer."

□   □   □

Unfortunately, the Du Pont official's reaction was about as interesting as they came in the summer following the release of the paper in *Nature*. Expecting a barrage from the media, University of California, Irvine, officials sent out a press release with quotes from Rowland that explained the situation very clearly.

"One of the troublesome aspects of the situation is the delayed action effect while the materials are diffusing upward," Rowland told a small group of reporters. "If the destruction of the ozone because of the chemicals ever becomes measurable, it will be too late to reverse and the problem will remain for decades."

Whether it was because Rowland and Molina weren't known as atmospheric chemists or because the implications of the theory seemed like something out of a bizarre science fiction novel, the media's reaction to the *Nature* paper was tepid. Several local newspaper reporters interviewed Rowland and Molina and Reuters wrote a brief story from the *Nature* paper. As far as Rowland could tell in the ensuing days, the story never got much beyond the borders of California. And while some overseas newspapers ran the Reuters piece, several of them reduced the story to a brief item, simply crediting the theory to "two scientists" in California. At that point, the "two scientists" certainly didn't seem destined for the science hall of fame.

Oh, well, thought Molina, who had spent the spring boning up on the atmosphere and learning how to do calculations on a computer. (He had churned out the original theory using paper, pencil, and calculator.) This was a rather esoteric problem. It consisted of invisible gas, invisible radiation, and a delayed reaction. It was no wonder the public had a hard time relating to it. With any luck, he thought, the scientific community will recognize its importance.

But reaction from the scientific community was equally slow to surface. Molina and Rowland might have expected this. It had taken *Nature* almost six months to have the paper reviewed by experts and to figure out what to do with it.

Although they felt the problem needed immediate attention, the lack of interest might be a good thing, the pair reasoned. It would give them a chance to do some more calculations before other scientists started trying to pick the thing apart. Johnston's warning—"Are you ready for the heat?"—had stuck with Rowland and Molina.

"We live on our reputation as scientists," Rowland told a reporter from *Orange County Illustrated* magazine. "We have to be very hesitant about putting our names on something unless we are sure about it. In this particular case we looked awfully hard for loopholes. Our conclusions were so far-reaching that we knew they would receive considerable attention, although I don't think either of us had any real concept of what that actually meant."

There was one exception to the scientific community's apparent lack of interest in the research. While Rowland was on sabbatical, Cicerone met up with an acquaintance at the National Bureau of Standards named David Garvin. Despite the fact that Cicerone and Stolarski's paper on chlorine had been rejected, word had gotten around about the work, and Garvin could see that the Michigan and Irvine researchers were interested in the same problem. Garvin had worked with Rowland at Princeton and had often sent Rowland some of his work on the rate constants of chemical reactions, including many that are important in the stratosphere. Rowland and Molina were able to use some of this work in the *Nature* paper and Rowland had sent Garvin a preprint of the paper. While encouraging Cicerone to contact Rowland, Garvin also suggested that Rowland send the paper to Cicerone. It was something of a sticky situation for Garvin. It was inappropriate, he knew, to reveal too much about the Rowland-Molina paper. One simply didn't pass on the results of another person's work before the work was published.

Cicerone had never met Rowland or Molina, but he wrote to Rowland, explaining his research and asking for information on the Irvine researchers' work. Rowland sent a copy of the paper to Cicerone who, on the first reading, was struck by the thoroughness, creativity,

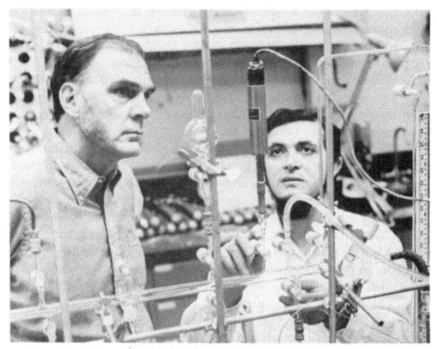

Chemists F. Sherwood Rowland and Mario Molina shown at the University of California, Irvine, in 1974, shortly after they discovered that manmade chemicals called chlorofluorocarbons could destroy the earth's ozone layer. (Photograph by Laurel Hungerford; courtesy of the University of California, Irvine.)

and, above all else, the implications of the work. One hundred space shuttle programs operating around the world couldn't match the damage these ubiquitous household chemicals could do to the ozone layer.

"It was quite obvious that what they had come up with was more important than what we had come up with on the space shuttle as a source of chlorine in the atmosphere," Cicerone later recalled. "It was clear to me that they had it right; that they had identified the most important part of the problem."

Cicerone and Stolarski met Rowland and Molina in July at a science meeting in San Diego. The paper had been published with few repercussions, and the three men talked at length about what to do next. With the paper having received so little attention, it seemed implausible that a couple of relatively unknown chemists could convince the world of this horrible threat. It would be nearly impossible, Cicerone suggested, to take the issue to the public and to policy makers in any convincing manner. The chemicals were ingrained in modern life. And yet, speaking out was the only responsible thing to do.

That opportunity arose when Rowland participated in a press conference after delivering his first major presentation of the research that fall at a meeting of the American Chemical Society. An ACS official had requested that he hold a press conference and Rowland agreed to cooperate. Cicerone and Rowland then decided that, after the ACS meeting, the Michigan researchers would issue a press release on a study that supported the Irvine findings.

"He decided to tell it the way he saw it, be honest, and hope that somehow people would be willing to listen to details and try to understand them," Cicerone remembered.

Rowland and Molina spent the summer watching the Watergate impeachment hearings and putting together a 150-page support paper on the theory in preparation for the September meeting of the American Chemical Society in Atlantic City.

The American Chemical Society is a huge conglomeration of chemists from industry, academia and government that holds national meetings twice a year and conducts several smaller regional meetings. The national meeting, which lasts a full week, features hundreds of presentations on new findings in chemistry. In 1974, it was the job of ACS news manager Dorothy Smith to read the abstracts of the papers, sent in advance of the meeting, and pull out those that might be of interest to the media. The ACS meeting, while failing to attract the science writers from some major metropolitan newspapers, always drew reporters from the wire services—United Press International and the Associated Press—and a few writers from the science journals and industry publications.

Smith was a savvy and determined woman who later earned a doctorate in chemistry and went to work at the National Academy of Sciences. By the time she left the ACS in the late 1970s she had also become known as someone who didn't feel it was her responsibility to protect the chemical industry as it came under increasing attack for its role in polluting the environment. (The attacks on the chemical industry at this time, she felt, were simply part of increasingly sophisticated instrumentation developed in the 1960s and 1970s that allowed scientists to assess, for the first time, how human activities could affect the environment.) Smith's job with the ACS was to see to it that the press was fed the most interesting stories, as long as those stories were based on good, legitimate chemistry. When she saw Rowland's abstract and the discussion of CFCs deadly potential, she was intrigued and called Rowland.

"Are you talking about Freon?" she asked.

Yes, said Rowland, explaining the theory in greater detail. Although the debate over the ozone issue was to rage for another decade, Smith detected something about Rowland that convinced her the issue was legitimate. "I always thought he had perspective," Smith said years later.

Smith proceeded to type up information about the upcoming press conferences and sent it out to the media. Not surprisingly, a Du Pont official saw the press release and put in a call to Smith.

"He told me this was an insignificant story," Smith recalled. "I got the impression that it was not good chemistry."

She also got the impression that Du Pont was very worried over this paper. Smith now needed to find some neutral party who could advise her objectively on the legitimacy of the Rowland-Molina paper. She finally reached some scientists at the National Oceanic and Atmospheric Administration, a government-run agency, who were familiar with the theory. They assured Smith that the work was good chemistry and raised valid questions, cementing her decision to go ahead with the press conference.

Before the ACS meeting adjourned, however, Smith was needled once again about her decision to hold a press conference on the CFCs-ozone theory. Heading out to lunch one day during the meeting, she encountered an ACS member who sat on the organization's public relations committee. One of the committee's responsibilities was to oversee the news bureau.

"I heard that story," the committee member called to Smith as he pushed in through a revolving door at the Atlantic City hotel and Smith pushed out. "Now why would you want to publicize that?" he chided her.

"A lot of people are interested in it," Smith called back, stepping onto the boardwalk and chuckling.

By the time Rowland presented the work in Atlantic City, the Irvine chemists' calculations on ozone loss were more pessimistic than they had originally stated. They estimated that if CFC production continued to increase at the present rate of 10 percent a year until 1990 and remain steady thereafter, the chemicals would cause a 5 to 7 percent ozone loss by 1995 and a 30 to 50 percent loss by 2050. Trying to place the information in a context that would demonstrate its importance, Rowland pointed out that skin cancers could be expected to rise drastically from the increase in ultraviolet light that would reach the earth if unusual amounts of ozone were depleted. He raised the possibility that climatic patterns could shift due to temperatures dropping in the stratosphere. Stepping outside the neutral world of science, Rowland and Molina then entered the world of politics, a world that Rowland, especially, was never to escape. They suggested that the atmosphere was not an inexhaustible sink into which human waste could be dumped and forgotten. They said that chlorofluorocarbons—an $8 billion dollar industry in the United States and the basis for human hygiene and housekeeping convenience the world over—should be banned.

□   □   □

While many participants of that ACS meeting thought the Rowland-Molina hypothesis was, as Rowland put it, "off the wall," the theory soon picked up steam and began to make headlines around the world. The issue had finally received some public attention—despite what Rowland and Molina later described as a hard lesson in the world of public relations.

Deciding to take full advantage of the press conference Smith had arranged, Rowland and Molina and some other symposium participants who had given papers on ozone agreed with Smith to organize the press conference in what they thought would be a logical, scientific format. First, others in the symposium group would provide reporters with a bit of background about ozone. Next, other researchers who had been doing measurements of chemicals in the atmosphere would speak. Then, as the suspense built, Rowland and Molina would explain their CFC-ozone depletion hypothesis.

But, as it turned out, the press had a lesson for the scientists: In the deadline-distorted world of competitive journalism, one does not withhold the biggest news until the end. Long before Rowland and Molina even got to their theory, a crestfallen Molina watched as some of the best-known reporters fled the room, obviously bored.

"That's not the way to do a press conference," Molina sighed, after the session, fretting that the reporters were still unaware of the theory.

Some reporters stayed until the end of the press conference, however, because news of the CFC-ozone theory, along with its drastic implications, began appearing in newspapers across the country within days. Some of the stories repeated the old doomsday theme, however, making light of the theory. Throughout the 1960s and early 1970s, the public had been hit with a barrage of "environmental crisis" stories. Every month, it seemed, someone was reporting a new threat to human health or the environment.

"Hissing Toward Doom," "Death to Ozone," "Aerosol Spray Cans May Hold Doomsday Threat," read the headlines in the fall of 1974. But many of the nation's best science writers gave the story the attention and credibility it deserved. And several things began working in the scientists' favor.

For one, Rowland and Molina were beginning to pick up some scientific support for the theory. At the ACS meeting in Atlantic City, scientists at the Naval Research Laboratory had reported that air samples from the Arctic, north of the Norway mainland, showed the presence of chlorofluorocarbons in both populated and remote areas. And, two weeks after the ACS meeting, Cicerone and Stolarski reported in the journal *Science* that their computer calculations predicted a 10 percent decrease in ozone by 1985 to 1990.

Cicerone and Stolarski's paper, which was actually reported in the popular press two weeks before it appeared in *Science*, gave Rowland and Molina a big boost of credibility. Although they weren't Nobel prize winners, Rowland, Molina, and Cicerone represented highly respected research universities. The three were articulate and, although they were new at speaking with the press, had respect for the nation's science writers and were willing to spend time explaining their research. "The press felt, frankly, that we were unusually helpful scientists," Cicerone said.

A few months after Cicerone and Stolarski's paper appeared, Harvard atmospheric scientists Michael McElroy and Steven Wofsy (who had been involved with research on the SSTs-chlorine issue) predicted that a 10 percent annual increase in the chemicals would decrease the ozone layer by 10 percent in 20 years and by 40 percent by 2014. And the Harvard researchers were even better than Rowland and Cicerone in the publicity department. The day before Cicerone and Stolarski's findings were published, McElroy, a trim man with red hair and an authoritative manner, had been quoted extensively by *The New York Times*' esteemed science writer Walter Sullivan about the Harvard research. The story had barely mentioned Rowland and Molina. Clearly, the competitive fires among scientists interested in the ozone issue had been stoked.

But other influential scientists were far from convinced that CFCs spelled disaster. James Lovelock, reflecting on his initial discovery of CFCs in the atmosphere over the Atlantic, observed, "The presence of these compounds constitutes no conceivable hazard."

Two weeks later, however, in an issue of *New Scientist* magazine, Lovelock noted that his previous comments had been written several months earlier, before he knew the details of the Rowland-Molina hypothesis. His initial assessment, he said, may have been too hasty. But he added, "It may be as unwise to assume that we are now at an early stage of a serious global pollution incident."

Speaking to a British newspaper reporter later that month, Lovelock chastised the American scientists for making such a big deal over CFCs. He criticized Rowland for "oversimplifying" the facts and called for a "bit of British caution."

"I respect Professor Rowland as a chemist, but I wish he wouldn't act like a missionary," the outspoken chemist said. "This is one of the more plausible of the doomsday theories, but it needs to be proved. The Americans tend to get into a wonderful state of panic over things like this. It's like the great panic over methyl mercury in fish. The Americans blamed industry until someone went to a museum and found a tuna fish from the last century with the same amount of methyl mercury in it."

That "someone" happened to be Sherry Rowland. In 1971, Irvine chemist Vincent Guinn, a forensic chemist, and Rowland had quieted the fears of fish-eating Americans by showing that tuna and swordfish naturally contained certain levels of mercury. Although that environmental flap was a minor one compared with the ozone issue, the Irvine group had, however briefly, become heroes to the fishing and food industries, much to the chagrin of some environmentalists.

The additional scientific backing from the Michigan and Harvard groups was enough to bring the CFC-ozone theory to the attention of the nation's policy makers. On October 8, the National Academy of Sciences, a government-sponsored organization, named a five-person panel, which included Rowland, to determine the seriousness of the problem and decide whether a full-scale investigation was in order.

The group met on October 26 in Washington and recommended that a one-year investigation begin immediately to determine the magnitude of the threat and assess the need for a ban of the chemicals.

Consumer groups and local governments also responded quickly to the CFC-ozone depletion theory. In Ann Arbor, Michigan, where the very vocal Cicerone had attracted a good deal of attention, the city council voted to urge citizens to follow a voluntary ban on the use of aerosol sprays containing CFCs. The council also agreed to consider a resolution for a mandatory ban.

And yet the Ann Arbor City Council knew it was acting symbolically on a problem that was obviously national and global. After the hearings, Mayor Jim Stephenson quietly contacted Congressman Marvin Esch, a Republican who represented the district that included Ann Arbor. Esch was a member of the U.S. House Committee on Public Health and the Environment. After speaking with Stephenson, Esch asked committee chairman Paul Rogers, a representative from Florida, to convene hearings on the CFC-ozone problem. Esch and Rogers then contacted Cicerone and asked him who should testify at the hearings. Cicerone suggested several scientists, including Rowland, Molina, and McElroy. This grass-roots action by a small town's city council culminated that December in the first federal hearings on this international problem.

Cicerone, who was deeply and emotionally committed to a CFC ban, spoke out on the dangers of the chemicals whenever possible. But his enthusiasm was not appreciated by the conservative University of Michigan administration. In one meeting before a Michigan dean, Cicerone was informed that he was causing trouble by raising new issues and bringing in new research money that would be difficult for the university to manage.

He was also invited to a meeting with the university's vice-president in charge of public relations to discuss his close relationship with the media. The purpose of the meeting, Cicerone suspected, was to make sure he "didn't have three heads." But Cicerone used the occasion to politely ask the administrator for some tips in dealing with the press.

"There are three rules," the administrator said. "Never talk to the *National Enquirer*. Never talk to the *National Enquirer*. Under no circumstances talk to the *National Enquirer*."

Cicerone left the man's office feeling quite sure he was on the right track.

Rowland, too, would have enjoyed seeing his home town respond to the ozone threat. But when a neighbor of Rowland's who was on the Newport Beach Planning Commission tried to get the Newport Beach City Council to consider the matter he was rebuffed and criticized for wasting the council's time on such a trivial matter.

While some policy makers considered the issue trivial, the theory galvanized the nation's environmentalists. Several enthusiastic environmental groups joined Rowland and Molina in calling for a ban of CFCs. In Detroit, a group called the Clean Air Movement urged Americans to boycott aerosol sprays containing CFC propellants. (Although CFCs were used in refrigeration, air conditioning, and other aspects of industry, the aerosol sprays were considered a frivolous, "nonessential" use of the chemicals.)

Much to the dismay of the CFC industry, consumer groups also noted that aerosol sprays weren't even economical. The aerosol sprays, they pointed out, contained from 50 to 90 percent propellant and only a small portion of the active ingredient. There were also some health threats associated with using aerosols. CFCs in high concentrations were known to cause cardiac arrest and, in smaller doses, were suspected of causing changes in normal heart rhythms. Other studies also questioned what breathing the contents of aerosol sprays could do to the lungs. The Clean Air Movement advised consumers to find substitutes for aerosol sprays.

"Instead of shaving cream use an old-fashioned bar cream or an electric shaver," one environmental group advised. "Instead of spray deodorant use roll-on, stick, or powder. Instead of spray paint use brush on. Instead of spray insecticides, use a compressed-air applicator or hand pump."

Thousands of conscientious Americans apparently responded to the pleas—following the lead of James Lovelock's family and Joan Rowland who, one year earlier, had gathered up every aerosol can containing CFCs in the Rowland household and tossed them out.

The consciousness raising seemed to be working. Although an economic recession might have been partially to blame, sales of aerosol sprays dropped 7 percent in November 1974.

On November 21, the Natural Resources Defense Council, a newly formed environmental group, joined the ozone debate. (The NRDC would, by the end of the decade, become one of the few organizations to stick with the issue when it was no longer popular.) The council, comprised of environmentalists with legal expertise, was best known at that time for a suit that led the Environmental Protection Agency to order a 60 percent reduction in the lead content of gasoline.

Lovelock described the group as "like Ralph Nader, but not as responsible." This time, the group petitioned the federal Consumer Product Safety Commission to ban nonessential uses of aerosols containing CFCs.

Dr. Karim Ahmed, a staff scientist at NRDC, had just joined the organization when word of the ozone-CFC theory began to circulate. Horrified at the environmental consequences the theory proposed, Ahmed called Rowland, Cicerone, and McElroy wanting to know more. He also spoke with Hal Johnston at Berkeley who believed that the theory was correct "beyond a reasonable doubt." After all, Johnston told Ahmed, much of the science that the ozone-CFC theory was based on had emerged from the CIAP studies. The database on chlorine chemistry and ozone was already there. After Ahmed's careful series of interviews, he and others at NRDC were convinced. This logic and lack of emotion became a mark of the organization's involvement in the issue over the next 15 years.

Rowland was clearly relieved at the show of support, telling his home town newspaper, *The Daily Pilot*, in late November, "I know of no knowledgeable scientist who feels that (a ban on nonessential uses of CFCs) should be put off for more than a year while we wait for additional facts."

But Rowland still harbored a deep-seated fear that the theory could be proved wrong. Almost a year after he and Molina had made their discovery, Rowland told a *Philadelphia Inquirer* reporter, "If someone finds the flaw, I hope it's not a stupid error."

□ □ □

As support for the theory broadened, so did a fierce counterattack from the CFC industry. Industry representatives hastily set up a $5 million project, managed by an international trade group called the Manufacturing Chemists Association, to fund research projects on potential CFC hazards. (Lovelock received one such industry-funded grant.) But, too ill-prepared at this point to fight in the scientific arena, industry representatives spent most of their energy following the ACS meeting in Atlantic City on their first line of defense—preserving profits.

The history of chlorofluorocarbons had been a glorious chapter in American industry, and no one was ready to part with the chemicals without a good fight.

CFCs were discovered one morning in 1928 by Du Pont chemist Thomas Midgley, Jr. Earlier, Midgley had spoken on the telephone with a colleague from General Motors. The colleague told Midgley that he and an engineer from the Frigidaire Division of General Motors had decided that the refrigeration industry was in dire need of new, better refrigerators. At the time, refrigeration was achieved through the use of ammonia or sulfur dioxide, and none of the chemicals worked too well. For one thing, the current class of refrigerants were highly toxic. Safety had been a major concern in the industry for some time. The three men discussed the ideal characteristics for a new refrigerant: It must be nontoxic, nonflammable, and stable. Something safe. Midgley, who also invented tetraethyl lead for gasoline, remarked that it was unlikely he could find one chemical to do the trick.

After some thought, Midgley decided that fluorine might be the right candidate for the refrigeration role. And, within weeks, he had come up with the compounds he was looking for. He called them fluorocarbons because they contained carbon and one or more of the class of halogen elements (fluorine, chlorine, bromine, and iodine). Because the chemicals later used in aerosol sprays contained mostly chlorine and fluorine, they were called chlorofluorocarbons. The chemicals are also called halocarbons. To demonstrate that this new class of chemicals was inflammable and nontoxic, Midgley even inhaled them and blew out a candle.

By October 1929, Frigidaire had a small fluorocarbon plant in operation and a new firm, Kinetic Chemicals, was formed to manufacture the chemicals. The company was owned by the giant chemical company, Du Pont, and GM, but Du Pont later bought out GM's interest and the plant became part of Du Pont's Freon Division. Freon was the trademark name Du Pont gave to the new class of fluorocarbon chemicals.

The future looked very bright for Du Pont. In the 1930s, the company adopted the slogan "Better Living Through Chemistry." The chemical age of American industry was off to an impressive start. The Freons named F-11 and F-12 were the most commonly used substances and turned out to be the most dangerous to ozone. Later, CFCs 113, 114, and 115 were also cited as particularly hazardous.

In the coming years five other American companies began producing the CFCs: Allied Chemical, Union Carbide, Pennwalt, Kaiser Industries, and Racon, and use of the chemicals soared. During World War II, carbon dioxide was used as a propellant for the dispersal of insecticides that protected the Allied Forces. The so-called "bug bombs" initiated the use of chemicals in aerosol cans. CFC-11, for example, became popular in the 1950s when mixed with CFC-12 as a propellant, a mixture that required an inexpensive valve because of the lower pressure requirements. Such a valve was invented by Detroit entrepreneur Robert Abplanalp and triggered the growth of the domestic aerosol industry after World War II. As propellants in aerosol sprays, the chemicals were everything industry dreamed of. They helped produce a fine spray. They were stable and didn't react with the product in the can.

While the makers of CFCs had a fortune at stake by 1974, so too did the people who used 75 percent of all CFCs—the people who packed the propellants into more than 300 aerosol spray products, distributed and marketed them. According to the Natural Resources Defense Council, aerosol containers in 1974 were used for cosmetics, perfumes, deodorants, window cleaners, air fresheners, insect sprays, lubricants, degreasing products, food containers, automobile products, and veterinary products.

How could modern people survive without them? What's more, cried industry officials, how could anyone suggest shutting down an entire American industry?

An editorial in the CFC industry trade journal, *Soap/Cosmetics/Chemical Specialties*, summed the situation up, "The press, television, radio and other segments of the media seem to be carrying on a vendetta against pressure packaging." The editorial urged packagers of aerosol sprays to stand by their spray cans. "Pressure packaging has served the world well and deserves a break."

One who was particularly miffed over a possible CFC ban was the king of the aerosol valve.

Robert Abplanalp had already earned the reputation of being one of America's great, bold entrepreneurs. After the war, Abplanalp, who operated a small machine shop in the Bronx, patented the leak-proof plastic and metal valve that allowed aerosol sprays to be mass produced. The valve transformed aerosol sprays from heavy steel containers of the "bug bomb" days into lightweight, aluminum cans. In 1947, the aerosol market consisted mainly of insecticides with sales totaling less than 5 million cans. In six years, production soared to 88 million cans. By 1973, more than 2 billion aerosol cans rolled off production lines, which meant the average American purchased about 14 aerosol cans a year.

The surge in aerosol uses had made Abplanalp a millionaire. His company, Precision Valve, earned $60 million in annual sales in 1974, according to a *Washington Post* report. Precision had plants operating in eight countries.

It had been a bad year for the outspoken, gregarious Abplanalp, however. He had suffered through the misfortunes of his close friend, Richard Nixon, and now people were attacking his beloved aerosols. Speaking in October before a meeting of the aerosol packagers, Abplanalp complained, "Extremists in the areas of ecology and consumer protection are today waging a more effective war on American industry than the most capable host of enemy saboteurs." Chlorofluorocarbons, he claimed, "would not make the sky fall in."

But no one had a bigger stake in the ozone issue than Du Pont, the largest of six U.S. manufacturers of CFCs. Du Pont's future with CFCs had looked quite rosy, despite the economic recession. In 1974, the company was putting the finishing touches on a new multimillion dollar CFC plant in Texas which was to become the largest manufacturing plant for the chemicals in the world. And company officials were finally starting to feel as if they had overcome one crisis with CFCs already.

In the early 1970s, manufacturers had become aware of increasing reports of young people abusing CFC products by sniffing the fumes to produce a state of intoxication. There were reports of deaths from the abuse and, in some communities, protests that the products were dangerous and should be banned. The situation greatly upset Du Pont officials. For decades, the company had been amazed at how perfect CFCs were. They had replaced dangerous refrigerants and survived years of safety tests. The company was proud of its reputation for safety. But now teenagers were abusing the product in a way that company officials felt they had limited control over.

In Du Pont's public affairs office, staff member Charles Booz was asked to put together a public information campaign to alert consumers to the dangers of sniffing CFCs. Booz spent many months with a group called the Aerosol Education Bureau working to get the message across.

"It had been astounding to us that the product was as perfect as it was," Booz later recalled. "It was absolutely amazing to us that a death could occur (from sniffing). We were very perplexed and concerned about our inability to stop something that we were not to blame for. We wanted to make the product available in a safe context. It was like telling everyone, 'Don't use drugs.' Finally, as the country became more sophisticated, you realized you can't really stop people from abusing a product."

The ozone-CFC matter was different, however. Suddenly, people were accusing Du Pont and other CFC manufacturers of reckless disregard for the environment. In the aftermath of the sniffing episode, this was a real bombshell. Du Pont officials couldn't believe their perfect product could, after decades, by found guilty of such a crime. Putting their best public affairs people on round-the-clock duty to answer media inquiries and sending researchers trained in public relations to scientific conferences to speak out on behalf of the product, Du Pont led the charge to defend CFCs and refute the Rowland-Molina theory. They were credited by many observers with handling the matter in the most professional manner of any of the CFC manufacturers.

"We welcome the scientific interest to develop hard, experimental facts about fluorocarbons and the atmosphere. We believe that when this data is in hand, it will exonerate fluorocarbons," a Du Pont official told the *Los Angeles Times*. But if the data were to prove otherwise, the official contended, "I doubt that we would continue to manufacture or sell a product that poses a hazard to life."

It was a position Du Pont was to maintain well into the next decade, although many observers felt that such data was available a full 12 years before Du Pont finally announced, in March 1988, that it planned to phase out all uses of CFCs.

In 1974, however, Du Pont officials swiftly pointed out that the theory was a hypothesis and that a major American industry should not be sacrificed when all the facts weren't in. Industries relying on CFC production generated $8 billion in business in 1974 and employed 200,000 people.

It was understandable that the chemical industry felt defensive about the Rowland-Molina theory. Du Pont was a responsible company and its chemists had, years earlier, been the first to question what became of CFCs after they were released into the atmosphere. According to Dotto and Schiff in *The Ozone War*, Du Pont had issued an invitation to other CFC manufacturers in 1972 to attend a seminar on the ecology of fluorocarbons. "It is prudent that we investigate any effects which the compounds may produce on plants or animals now or in the future," the invitation read. Industry chemists did, in fact, employ several scientists to conduct studies on CFCs, but these efforts were focused on the lower atmosphere. By 1974, just before the Rowland-Molina paper was published, the industry-sponsored scientists completed their studies, concluding that the chemicals were of no threat in the lower atmosphere. They had neglected to look any higher.

☐ ☐ ☐

December 11, 1974, dawned cold and clear in Washington. Dressed in a trenchcoat, a smiling John Ehrlichman showed up on Capitol Hill to testify on his role in the Watergate cover-up. In the Oval Office, President Ford talked of his plans to move ahead on peace settlements in the Middle East.

And, in the crowded hearing room, the first of many hearings on the ozone issue were called to order by Representative Rogers of the House Subcommittee on Public Health and the Environment. The hearings had been organized to call attention to a bill, introduced by Rogers and Esch, amending the Clean Air Act. Rogers and Esch, who were acting in response to Cicerone's concerns, proposed that the National Academy of Sciences conduct a full-scale investigation into CFCs and authorize the Environmental Protection Agency to ban the chemicals if they proved to be hazardous to human health or the environment.

"The entire matter rings of a science fiction tale, one we have all heard: how a planet, now barren, was destroyed by its very inhabitants," said Rogers, dramatically, calling the hearings to order. "Had not the evidence been brought forth by such reputable men of science, it would seem like bitter black humor—that the earth may be endangered and the villains of the situation are billions of aerosol cans."

The following testimony left even the representatives who had earned A's in college chemistry with tension headaches, so wide were the opinions of the threat.

Rowland was testifying before a political body for the first time in his career. In his deep and resonant voice—which was well suited to public debate and capturing the attention of an audience—Rowland summarized the issue by sounding a theme he would carry for the next 15 years.

"The fundamental problem is simply this—how long should we wait for someone to find the missing factor which might serve as a basis for an alternative scientific hypothesis before we act upon the conclusions which we readily derive from the only hypothesis available now?" he said.

Harvard's McElroy was more conservative. "It's not a matter of doomsday tomorrow. But if we wait too long the damage to our atmosphere maybe so great that we'll have a difficult time repairing it."

Dr. Lester Machta of the National Oceanic and Atmospheric Administration warned the committee not to expect science to come up with conclusive, irrefutable evidence. Science, Machta noted, just didn't work that way.

"Even after our best efforts, the issue may still remain unsolved," he said.

Industry representatives cautioned the committee to consider the economic impact the ban could have—a ban, they pointed out, that was based on an unproven theory. Said Dr. Igor Sobolev of Kaiser Aluminum & Chemical: "We may be about to extrapolate an unproven speculation, one open to serious question, into conclusions and laws that could disrupt our economy and indeed our way of life."

Raymond McCarthy, Du Pont's manager of the Freon Products Division, testified that the theory was speculative and that regulations were "unwarranted at this time." But he added a comment that Rowland took to heart and would later repeat to anyone who would listen. McCarthy conceded that if Freon proved to be a threat to the ozone shield, Du Pont "would stop its manufacture" of the chemicals.

To Rowland and Molina, that proof was already in the making. But the issue was not destined for a quick resolve. The 93rd Congress expired days after the hearings. On a policy level, nothing was accomplished other than to alert the American public and lawmakers that the issue required immediate attention. Rowland and Molina couldn't have agreed more. But interest in preserving the earth's precious ozone layer waxed and waned over the next 15 years. What effects this long delay might have had on the atmosphere are only now beginning to be addressed.

3
The Politics of Science
January 1975—June 1975

They could cope with war. They knew how to handle economic recession, inflation, and government corruption. But when it came time to
decide what to do about large-scale attack from aerosol spray cans,
the United States government was stumped.

The choice was not easy: Should the nation risk economic peril
to act on a theory steeped in scientific uncertainty and whose consequences, even if true, might not be felt until late in the next century?
Or was it better to wait until scientific certainty had been established
and take the risk that the delay might lead to irreparable environmental
damage that future generations would have to contend with?

Although the Rowland-Molina hypothesis had attracted widespread attention by early 1975, policy makers were determined to wait
for scientists to come up with a more complete picture of just how
the world would be destroyed by spray cans before doing anything
about it. An attorney for an aerosol company described the situation
for a *New York Times* reporter: "Legend tells us that Henny Penny,
who first postulated that the sky was falling when an acorn fell on her
head, was subsequently proved wrong."

□   □   □

Deciding what action, if any, the federal government should take fell
to a group called the "Committee on the Inadvertent Modification of
the Stratosphere." With a name like that, the CFC industry might have
expected the worst.

39

IMOS was organized by the federal government early in 1975 to assess the CFC-ozone issue. As a presidential task force, however, IMOS was proof that the White House had taken notice of the issue. The task force was chaired by Carroll Pegler Bastian and Warren Muir, whom several scientists later asserted were remarkably effective leaders. Featuring representatives from almost every federal agency, the task force held a dramatic one-day hearing on February 27 in the New Federal Office Building in Washington. Although its burden was a heavy one, the committee was reassured by Harvard's Mike McElroy that enough brains were available in the atmospheric science community to make sense of the matter.

"For once the scientific method is working," McElroy told the committee. "The scientific community is focusing on potential scientific problems and bringing them to your attention in time to do something."

McElroy, a trim redhead who speaks with the kind of distinguished tones that seem to fit the image of a Harvard academician, was an opinionated and competitive player in the ozone issue. But he seemed to be riding the fence regarding the severity of the problem and proceeded to inform the committee that the theory was in doubt. This provoked a rebuttal by Rowland and prompted a debate between the scientists that was just below the boiling point. The disagreement served to illustrate for lawmakers that they would have to make a decision on a scientific issue that was steeped in uncertainties. The bottom-line question was whether the CFC-ozone problem was serious enough to warrant government regulations. Another chemist testifying before the committee summed the situation up: "The disagreement is not scientific, but one of public policy. Scientists must provide the best information and define the alternatives."

The lack of agreement between Rowland and McElroy wasn't the only thing to get the IMOS committee off to a rocky start. The matter of who had jurisdiction over CFCs was causing a lot of confusion. The several federal agencies, represented by their legal counselors, each stated that their agencies had no jurisdiction over aerosol spray cans. Listening to the committee wrestle with the issue, Cicerone grew more depressed.

"This is going to be an impossible job," Cicerone sighed.

But the IMOS committee proved to be a catalyst for the federal government's attention to the matter over the next three years. Despite the shaky start, Bastian and Muir hit the streets of Washington in an attempt to sort out each agency's responsibilities and to assign juris-diction so that federal legislation regarding CFCs could function. The committee's conclusions, which were released in June, clearly gave the issue new importance. Calling the issue a "legitimate cause for concern," the committee reported that CFC restrictions might be nec-essary. It then handed the ball over to the National Academy of Sciences saying that if the NAS found that CFCs were hazardous then the federal government should act to restrict their uses. (This statement came as a bit of a shock to the NAS, which now found itself with the sole responsibility for ruling on the matter.) The committee also suggested that aerosols containing CFCs be labeled as such and recommended that Congress pass the Toxic Substances Control Act which would give the EPA authority to control products deemed hazardous. It recom-mended that the State Department alert other countries to the possible dangers of CFC use and called for a direct exchange of information with the Soviet Union.

While President Ford's top science advisers said the evidence was still not strong enough for an immediate ban, the IMOS report seemed to convince many government officials that regulations would prob-ably be necessary.

"We are going to minimize the use of aerosol cans using fluo-rocarbon compounds while studies are completed," Dr. Russell W. Peterson, chairman of the White House Council on Environmental Quality, told the *Washington Post*. Peterson predicted that a ban would be in place by January 1978. Peterson's words were of interest to many observers. He had worked for Du Pont for more than two decades before entering politics and becoming governor of Delaware.

The IMOS report dismayed the CFC industry, although it re-sponded with its characteristic vigor. Du Pont research scientist Ri-chard Ward told the *Los Angeles Times* that the report "essentially concurs with the industry's position that there is no appreciable danger in continued use of the fluorocarbon compounds while studies are completed."

Even some nonindustry observers considered the IMOS decision premature.

"Hysteria Ousts Science from Halocarbon Controversy," read a headline in the British science journal *New Scientist*. The story featured an interview with James Lovelock who said that the American political reaction meant that "scientific arguments no longer count."

□ □ □

The CFC-ozone controversy was now in the hands of the National Academy of Sciences. It was the NAS who had first recommended that the government fund a complete study of the issue. NAS officials were hesitant about handling this political hot potato, however, and for several months kept a low profile regarding the issue. Finally, in April 1975, almost six months after agreeing to organize a formal CFC-ozone study, the academy named a 12-person panel to assess the problem.

The panel was divided into two task forces. One would review the scientific evidence and pronounce judgment on the validity of the Rowland-Molina hypothesis. The other task force would address public policy issues and make recommendations about what should be done. The panel was instructed to deliver its report in one year: by April 1976. In the meantime, the panel, which was generally comprised of people who had not yet taken a firm stand on the issue, was instructed to keep its deliberations quiet. But it became apparent rather quickly that this matter could not be confined to the chambers of the NAS. There was just no way to avoid the politicking that surrounded the controversy, and committee members were pressured by the various interest groups—especially industry—to hear them out.

"It was hard to insulate ourselves from that," Harold Schiff, the Canadian chemist and panel member, later recalled. "What we wanted to do was hear all sides. And we had representation from industry. But we were also getting pressured to hear their side. Some of that pressure we thought pretty unfair."

Not everyone was thrilled that the matter would be decided by the NAS. Called the "headquarters of the politics of science" or the "supreme court of science" by some, critics questioned the academy's ability to deal with this delicate matter. The academy's history did little for its image as a fair and reasonable "science court."

Chartered by Congress on March 3, 1863, the organization's role was to "investigate, examine, experiment, and report upon any subject of science or art" when requested by the federal government. Some of the NAS's first tasks were to help the military improve on the accuracy of compasses used on iron ships during the Civil War and to show the government how to test for purity in whiskey.

In the late 1960s, however, the NAS had suffered a test of credibility by repeatedly insisting that SSTs would not harm the environment. The academy's position was so unpopular that its own members were up in arms. In his 1976 book, *The Brain Bank of America*, author

Philip Boffey described the academy as a puppet of government. Others, too, criticized the organization for not being an independent voice in science. In *The Ozone War*, Schiff described Hal Johnston as among those most in doubt of the academy's ability to handle the CFC-ozone matter. In a letter to the academy, Johnston suggested that the people deciding the matter for the academy were gullible and could easily be dissuaded by the public relations genius of the chemical industry. Johnston also criticized the panel for its lack of expertise in atmospheric chemistry. He demanded to know whether the NAS panel would make use of the three-year findings of the Climatic Impact Assessment Program.

Johnston had good reason to doubt whether the committee would be fair. There had been a great deal of skepticism by academy members regarding Johnston's theory on SSTs and ozone depletion, and Johnston hadn't forgotten that. Academy members saw Johnston as too emotionally involved in the SST debate and lacking objectivity. Rowland could have suffered a similar prejudgment by the NAS, but most NAS panelists felt the CFC issue was presented in a different light.

"Sherry didn't make that mistake of becoming too emotional," Schiff said. "Sherry did a careful study with Molina and didn't come out making extravagant claims right at the beginning. So it looked like a fairly sound case."

Rowland was satisfied with the panel, feeling the group was represented by experts in the field and with people known for their ability to be analytical. Cicerone also felt that the committee members respected Rowland and would be conscientious and probing. But he had a nagging fear that a sizable dose of skepticism had already pervaded the committee. The NAS had a reputation for being conservative, and it was not out of the question that they would carefully consider industry's immense economic and political power.

And there was still the jurisdiction problem. Even if the NAS found the issue to be worthy of government action, who would decide what regulatory action was needed and enforce it?

No one government agency could claim authority over the use of CFCs. The Environmental Protection Agency regulated CFCs in pesticides. The Consumer Product Safety Commission ruled over the use of CFCs in home refrigerators, furniture polish, and window cleaners. The Food and Drug Administration regulated the chemicals used in cosmetics and drug dispensers.

A possible legislative solution to the bureaucratic entanglement was proposed in the spring of 1975 by Congressmen Rogers and Esch.

Their bill sought to amend the Clean Air Act, giving the EPA authority to prohibit the manufacturing and sales of all CFCs should they prove to be hazardous to human health or the environment. Similar legislation was introduced in the Senate with the backing of Dale Bumpers of Arkansas, Bob Packwood of Oregon, and Pete Domenici of New Mexico.

Rogers and Esch viewed the CFC-ozone issue as a vehicle to bring about additional regulatory framework to address environmental matters. And they weren't the only politicians to try and capitalize on the CFC-ozone issue. By that fall, 16 bills had been introduced in Congress that could affect aerosol production. Among them was a bill by Utah Senator Frank Moss to furnish NASA with funds to conduct intensive research on the upper atmosphere. (NASA, indeed, was to become the primary government agency to oversee studies on the once-ignored stratosphere.) Another bill, submitted by Congressman Les Aspin of Wisconsin, was similar to the Rogers-Esch bill but contained a provision for issuing licenses to manufacture or import CFCs.

In hindsight, the flurry of legislative action that swelled around the CFC-ozone action may have been a response to the post-Watergate aura that permeated Washington. With trust and confidence in the executive branch of government at an all-time low, House and Senate representatives were grabbing power and authority wherever possible.

Environmental committees, such as Rogers' and Bumpers,' delved into substantive issues in a manner that, many observers felt, had never been seen before or has been witnessed since. That was especially true of the 1975–1976 series of hearings on the matter under the Subcommittee on the Upper Atmosphere, chaired by Bumpers. An arm of the Aeronautics and Space Science Committee, Bumpers' committee had no legislative authority but was given the responsibility of overseeing everything that happened approximately 10 miles above the earth's surface. Bumpers took the responsibility very seriously, latching onto the CFC-ozone theory in early 1975.

The very fact that Bumpers had been given the chairmanship of the subcommittee was testimony to his ambition. He was part of a freshman class of senators who had demanded subcommittee chairmanships from the Senate leadership. The Subcommittee on the Upper Atmosphere was a small one, but Bumpers' enthusiasm along with Senator Domenici's interest in the problem provided a climate for sustained discussions on the CFC-ozone problem.

In addition, Bumpers was assisted by a skilled and savvy legislative assistant named Bill Moomaw. Moomaw was a nice fellow with an objective, inquisitive air about him. He had a Ph.D. in physical chemistry and had arrived in Washington that spring as a congressional science fellow under the sponsorship of the American Association for the Advancement of Science. There were (and still are) very few technically trained people on any congressional staff, and when Bumpers discovered that Moomaw had expertise in photochemistry, he offered the scientist a position on his staff. Throughout the hearings, Moomaw regularly advised Bumpers of the latest scientific developments in the theory, keeping the subcommittee at the forefront of the issue. Years after the Bumpers' hearings—which produced three 500-page volumes of testimony—the scientists who came before the committee praised the hearings as among the most comprehensive and scientifically thorough they had ever experienced.

The Bumpers' hearings, like almost every discussion on the theory, were never boring. Early on, committee members were forced to deal with charges by industry that there were no measurements of CFCs in the stratosphere or depletion of ozone. "We don't even know if those things are there," one Du Pont representative charged at the hearings. The witnesses couldn't even agree on the value of a scientific theory, Moomaw recalled.

"It was interesting to me as a scientist to hear (industry representatives) refer to it disparagingly as 'the theory.' To a scientist, a theory is sort of the pinnacle of intellectual accomplishment. To industry, theory meant nothing more than your speculation versus my speculation."

Some subcommittee members, however, were suspicious of the scientists. While the scientists traveled to Washington feeling it was their duty to explain the issue from a scientific standpoint, some subcommittee members and staffers wondered privately whether the scientists' motives were pure or whether they were just another special interest group trying to drum up money for research.

A hallmark of the early CFC-ozone hearings was the range of opinions that existed among scientists—even scientists who were, ostensibly, on the same side. In trying to keep the committee completely informed, the scientists sometimes exaggerated their presentations. At one point, for example, it looked to perplexed subcommittee members as if everything could destroy ozone. While Rowland and Molina saw CFCs as the biggest threat, McElroy insisted that bromine, a compound related to chlorine that is commonly used in fire extinguishers, was as important a threat. (McElroy had the distinction of having his theory reported in the *National Enquirer*, which speculated that bromine gas containers could be launched into the atmosphere to destroy ozone like a "doomsday weapon that would cause equal harm to friend and foe alike.") The debate on nitric oxide from SSTs was still being mentioned as a potential threat to ozone, and McElroy also suggested that nitrous oxide released from fertilizers might be a problem. Other scientists questioned the damage nuclear weapons, which would release huge amounts of nitric oxide, could do to the ozone layer.

Bumpers managed to keep the discussion centered on the CFC-ozone theory, however, and often gave the scientists a run for their money.

A trial lawyer, Bumpers had a polite but forceful, almost tricky, way of eliciting responses from otherwise reluctant people. From scientists, he wanted the science explained to him in detail. From industry representatives, he wanted a statement of their true intentions. In one skillful interchange with Du Pont officials Roy Schuyler and Ray McCarthy, Bumpers pried forth a guarantee that was long remembered.

Bumpers: "Mr. Schuyler, let me ask you if the refrigerants used prior to the use of Freon—were they flammable?"

Schuyler: "Some were. Ammonia, for example."

Bumpers: "And toxic?"

Schuyler: "Certainly SO_2 is toxic. Methylchloride could be considered to be toxic."

Bumpers: "Was Du Pont or the industry ever mandated by the Government to discontinue the use of toxic or flammable refrigerants?"

Schuyler: "So far as I know, no."

Bumpers: "This is something the industry did on its own?"

Schuyler: "Freon immediately took over because of its value in use against the other materials then in vogue."

Bumpers: "You stated you are willing to stop production of Freons, if the link to the destruction of the ozone can be established. What level of proof do you think would be necessary?"

Schuyler: "It is very difficult to define, Mr. Chairman, but certainly if it were not refutable, and if we could look at the data and the data were reasonable and right and could be sustained, certainly we would withdraw it. Perhaps Ray McCarthy has a word. Do you have anything to add to that?"

McCarthy: "No; I do not believe so."

By the fall of 1975, it was no secret that Bumpers was impressed with the Rowland-Molina theory and the testimony of Rowland, Cicerone, and other supporters. Announcing that he and Domenici were designing legislation to regulate CFCs, Bumpers told reporters, "Nobody has come before the committee and denied a single facet of the Rowland-Molina theory."

Eventually, Bumpers joined Bob Packwood of Oregon in sponsoring an amendment to the Clean Air Act to ban the manufacture of aerosol spray cans after January 1, 1978, unless the EPA determined there was no significant environmental hazard. The amendment received only 28 votes, however, infuriating Bumpers. Another amendment to the Clean Air Act giving EPA authority to regulate the chemicals if they were found to be dangerous was introduced by Domenici, who was a member of the Public Works Committee. The Senate Commerce Committee also introduced similar legislation, which passed after much haggling in 1976.

□ □ □

Despite the dramatic arena afforded by Bumpers' hearings, it was clear that no government action would be taken until the spring of 1976 when the NAS report was released. But this foot dragging by the federal government didn't stop several state and community leaders across the country from demonstrating the power of grass roots political action. The concern over possible damage to the ozone layer shown by various state governments was not only gratifying to Rowland and Cicerone but was highly important in getting their message to the people who really mattered: consumers.

By the spring of 1975, 11 states had legislative proposals regarding CFC regulations, and the battle between the chemical industry and proregulation forces moved into the statehouses of Oregon, New York, and California. No one took the state measures lightly. The Council on Atmospheric Sciences, an industry group organized to lobby for CFC protection, feared state legislation could snowball into national legislation. At Du Pont, company officials viewed the movement to ban CFCs among state legislators as typical of the "little guys" being "opportunistic," of wanting to be seen as leaders on a national issue and "wanting to write a new law to solve the world's ills," as Du Pont spokesman Charles Booz put it. State legislation was silly, charged an editorial writer in *Chemical Week*. "How can what Minnesota does

Who's Who in the CFC-Ozone Debate

American Chemical Society (ACS): Scientific group organized to exchange information and promote chemistry.

Climatic Impact Assessment Program (CIAP): Study organized by the federal government in the early 1970s to assess the threat of supersonic aircraft on the atmosphere.

Committee on the Inadvertent Modification of the Stratosphere (IMOS): Organized by the federal government in 1975 to assess the CFC-ozone theory and decide what to do about it.

National Academy of Sciences (NAS): Agency that conducts scientific studies at the request of the federal government.

Natural Resources Defense Council (NRDC): Environmental group formed in the early 1970s and lobbied for ozone protection.

Council on Atmospheric Sciences (COAS): Industry group formed to lobby for CFC protection in the mid-1970s.

Manufacturing Chemists Association: Chemical industry trade group that sponsored studies on the CFC-ozone issue. Later became the Chemical Manufacturers Association.

Chemical Specialties Manufacturers Association: Industry trade group that included many aerosol businesses. Lobbied for CFC protection.

Organizations from the chemical industry, consumer groups, and government panels were active players in the CFC-ozone debate throughout the mid-1970s.

control what happens to the planet?" And yet, scientists, industry, and environmental groups flocked to the state hearings prepared to fight. The NRDC's Ahmed viewed the state hearings as extremely important but fretted that witnesses on both sides seemed to view state legislators as simple-minded folks who could be easily confused by the science.

"Everybody had an argument to try and confuse the matter," Ahmed said.

☐ ☐ ☐

Oregon lawmakers were the first to find out that this CFC-ozone issue was a game of hardball. The first state to seriously consider a ban on the sale or manufacture of CFCs, Oregon officials were unprepared to see their legislature turned into a stage for a scientific debate of global importance. The principal players for industry and the environmental groups eagerly attended the hearing at the request of the Oregon legislators to help lawmakers resolve the issue. After all, this was a first crucial test of whether the theory could gain popular support.

Rowland, too, thought Oregon might be a turning point in the public's recognition of the CFC problem. Although he was a relative novice at legislative matters (he had testified for the first time before Rogers' committee in December 1974), Rowland dutifully flew to Salem along with Molina. While one of them could have explained the theory, this was politics. And they had been advised by people who were more politically savvy that having both of them testify on the dangers of CFCs would double the effect.

The hearings were held on a Monday evening in May. Those testifying on behalf of CFC regulations agreed that Rowland should appear later in the evening in order to rebut anything their opponents had said. Although Hal Johnston was quite effective, telling lawmakers that the ozone layer was "the Achilles heel of our robust atmosphere," by the time Rowland was called to testify, the clock was approaching 11:00 P.M., the lawmakers' eyes were glazed, and the local newspaper reporters had long since fled to file their stories. The next day's news accounts would have to do without Rowland's succinct quotes.

"A benumbed committee sat through 4½ hours of testimony, with each member growing more bewildered by the scientific testimony of each successive witness," noted a reporter for the Eugene, Oregon, *Register-Guard*, quoting a committee member who said that a decision on whether to ban the chemicals would be made on "emotion instead of facts."

But one key committee member was persuaded to back the bill, and in late June, Oregon became the first state to ban the sale of CFCs in aerosols. Admitting that not all the scientific evidence was in, Governor Robert Straub signed the bill saying he would rather "err on the side of caution." The law, which took effect March 1, 1977, stipulated that shopkeepers selling CFC products could receive a $10,000 fine, a year in prison, or both.

New York became the second state to enact legislation on CFCs, voting later that year to require manufacturers of aerosol spray cans to place labels on the containers stating that use of the product could harm the environment. The New York law also gave the State Environmental Conservation Commission the authority to ban the sale of aerosols containing CFCs if the products were proven to harm the environment.

The New York state hearings were among the most lively of the early CFC-ozone debates. According to one account in *The New York Times*, industry hired two of the highest paid lobbyists in the state to work on the issue. And, as was the case in Oregon, the top draws from both sides showed up to defend their opinions. In one exchange, Rowland debated Du Pont's McCarthy in tones that were polite but heated. And, in another row, Ahmed squared off with officials from the cosmetic industry.

Ahmed, a normally calm and mild person, had become increasingly frustrated with what he called industry's "red herring" arguments. On several occasions, for example, industry representatives had stated that chlorine drifted into the atmosphere constantly from natural sources—such as sodium chloride evaporating from sea salt spray—and that the environment had long proven capable of handling this chlorine. Furious upon hearing this argument again at the New York state hearings, Ahmed attacked his opponents, ridiculing industry representatives for their lack of scientific logic. This small, background loss of ozone from natural sources was nothing new, Ahmed argued. But when manmade chemicals were suddenly added to this equation, the earth's ability to replace ozone as fast as it is destroyed is thrown off. After the session, Ahmed was approached by angry industry officials.

"You certainly tarred us with the same dark brush, didn't you?" one industry representative asked him.

"You people deserve it," Ahmed responded. "I've been going every place and listening to the same sad arguments."

The NRDC struck back at the New York hearings, using the occasion to alert consumers that aerosols were often not as economical as other products. NRDC representatives presented studies that showed how stick deodorants gave consumers more for their money than aerosol spray deodorants. Much to the dismay of industry, the information was entered into the public record.

□   □   □

Industry lobbyists persuaded several other states to vote down similar legislation. And CFC manufacturers and users were able to obtain some protection from within the federal government.

In July 1975, the Consumer Product Safety Commission turned down the Natural Resources Defense Council's petition to ban CFCs, saying there was insufficient evidence that CFCs harmed the atmosphere. Ahmed, who had pushed the petition, was disappointed. But he suspected that the fight had just begun. "This won't be the last word on it," he told fellow staff members.

The Department of Commerce also became an early ally of the chemical industry by discouraging CFC regulations. The agency pointed out that five industrial sectors could be directly impacted by federal or state restrictions: CFC manufacturing, air conditioning and refrigeration manufacturing and service, aerosol formulation, aerosol containers and valve manufacturing, and plastics manufacturing. According to Department of Commerce statistics, CFC production in 1974 reached $500 million. Almost 600,000 jobs with a payroll of $6.7 billion directly depended on the industry. Another 1.5 million workers were indirectly dependent on CFC production.

And it was obvious to federal lawmakers—as it was to everyone— that if the chemicals did harm the environment, it was hardly a matter to be handled by the United States alone. Although the United States was the largest producer of CFCs, the unique nature of the problem meant that the use of deodorants in England could affect the skies over Australia, while refrigerants used in America could cause ozone thinning over Antarctica. EPA chief Russell Train proposed that all nations cooperate in establishing worldwide guidelines to avoid environmental disaster.

"The world community can no longer duck this issue," said Train. "Increasingly, international cooperation on safeguarding our life support systems is an urgent necessity. The alternative is to sit back and wait for global disaster to overtake us."

Despite warnings of increased ultraviolet light and climate change, Train's recommendations went unheeded and international cooperation on the ozone issue lagged long after the scientific evidence confirmed the threat. Train's proposal, it seemed, was indeed too radical. Obtaining international agreement on the CFC-ozone issue proved to be several times harder than the domestic struggle. And industry, focusing on the need for international cooperation, stymied domestic regulations considerably by demanding that other nations become involved as a first step.

Other agencies, however, accepted the United States' role as one of leadership and pressed for domestic CFC regulations early.

In 1975, the EPA encouraged the pesticides industry to begin looking for substitutes for the chemicals. The FDA, too, warned the CFC industry that it was looking closely at the matter. And the Council on Environmental Quality urged that a ban on the nonessential uses of CFCs be implemented by 1978 unless evidence compiled by the NAS showed the chemicals did not pose a threat to the environment.

For the NAS committee members who were to decide the fate of CFCs, the pressure was mounting.

4
Innocent until Proven Guilty
April 1975—December 1975

If the Rowland-Molina theory had at first seemed like a joke to the CFC industry, the IMOS report had revealed that the issue was no laughing matter.

Industry insiders realized they would have to move quickly to preserve their business—or at least buy time to decide what to do. The battle was fought on two fronts. The first was the scientific arena which was the most important but least effective of industry's armaments. The second was the public relations arena which was alternately effective, bumbling, and amusing.

One could argue about how convenient CFC products were and how many jobs depended on the industry, but both industry and environmentalists knew the real issue centered on proving or disproving the Rowland-Molina hypothesis. No one could say just what degree of proof would be needed for the government to ban CFCs.

The emotional nature of the conflict became apparent at the 169th national meeting of the American Chemical Society at the Barclay Hotel in Philadelphia in April 1975. Sessions on ozone depletion and CFCs ran all day, broke for dinner, and resumed at 8:00 P.M. for hours of additional debate. Between sessions, groups of scientists hovered together, exchanging gossip or the latest calculations on ozone loss. According to several accounts, scientific decorum was happily misplaced. Some industry representatives described the meeting as "hostile." Said one writer for *Aerosol Age*, "At some of the hearings or meetings, the researchers react with a good deal less than calm. In fact, they ride in on their white horses preceded by their public relations releases as if they were running for public office."

Rowland found the meeting no less scientific than other ACS symposia. Still, some observers questioned the motives of everyone involved in the ozone debate—as if million-dollar profits and environmental catastrophe weren't enough reason to be concerned. In one industry editorial published after the meeting, the writer complained that, "The whole area of research grants and the competition among scientists to get them must be considered a factor in the politics of ozone."

Even Lester Machta, director of Air Resources Laboratories at NOAA, observed, "It's difficult to get behind the true motivation of some of the scientists involved." But, he added, "I think many of them have the feeling they wish they were wrong so they would not upset the economy."

The intense media coverage played a part in stoking the fires between scientists and industry. Privately, the two factions showed measured respect for each other—particularly among the scientists. Early in the debate, Cicerone found strong resentment and disbelief among industry people whose jobs were in marketing. They simply couldn't understand how these chemicals, which had been around for so long, could now come under attack. Many blamed the media for exaggerating the issue. As for industry scientists, they were accustomed to letting their marketing and public relations officials do the talking. But there were nights, over drinks in hotel bars, when industry scientists seemed as concerned as anyone.

"Look, I'm a person too," they would tell Cicerone. "I have children. I want to have grandchildren. We're interested in the future, too."

Some industry officials tried to keep the fight from becoming personal.

"I don't think we have a fight with anyone," Du Pont public relations representative Jim Reynolds told the *Philadelphia Inquirer*. "We have a disagreement with Rowland, Molina, McElroy, and Cicerone. They're good, honest men and this is really something they believe is valid and we just as honestly and sincerely believe is invalid."

Nevertheless, Dorothy Smith bit her nails as the ACS press conference on the CFC-ozone issue neared. The mood of the meeting had been unusually cantankerous and she knew the press conference would be antagonistic. Smith remembered the last ACS press conference on the Rowland-Molina theory when Du Pont scientists had interrupted the speakers and used the press conference for their own forum. She feared a repeat performance.

"This isn't a normal press conference," Smith told a colleague. "A press conference is for the press. Half of this audience is going to be scientists."

Attempting to maintain some control over the session, Smith intervened when industry officials began raising their hands to ask questions. She demanded that they state their names and affiliations before asking questions, a tactic she hoped would reduce the number of combative, nonmedia inquiries.

Rowland and Molina were becoming more adept at handling media questions and industry punches with every new encounter. Molina, although soft spoken and younger than most of the scientists involved in the debate, planted his feet and never failed to meet a challenge. He had spent most of the past year boning up on the finer details of atmospheric chemistry and finding support for the theory. As long as he was armed with his facts and figures, he felt confident enough. And, when all else failed, Molina would try a little gentle humor. He often explained that a ban on CFCs in aerosol spray cans would not be a particularly difficult first step in addressing this environmental threat. After all, Molina explained, products now packaged in aerosol sprays could be put in pump dispensers that didn't require CFCs. Once, when asked by an industry spokesperson if he had considered the fact that pump sprays might be more dangerous than CFC-containing aerosols, Molina looked puzzled and then shrugged. "Pumps are not particularly dangerous unless you throw them at people or swallow their contents," he said with a straight face.

Most of the media pressure, however, was put on Rowland, the more experienced and articulate of the two.

Rowland made himself available to the press with unfailing patience, sometimes spending several hours a day talking to reporters from his campus office. By being accessible to the media and knowing how to speak in reporters' language, Rowland clearly furthered his own cause—to seek a ban on nonessential CFC uses. (When it came to publicity, however, he knew where to draw the line. In 1977, a movie studio released a film, entitled *Day of the Animals*, about destruction of the ozone layer and animals running amok because of the increased ultraviolet light. While Cicerone turned down repeated invitations from the producers to appear on television with actress Linda Day George, Rowland passed on requests to promote the film.)

But his media relations skills brought him criticism. Robert Abplanalp, for one, accused Rowland and Molina of having "chosen to run to the media." Abplanalp saw Rowland's advocacy as inappropriate for a university scientist. The frustrated industrialist even wrote to Daniel Aldrich, the chancellor of the University of California, Irvine, and the man who had hired Rowland, complaining of Rowland's activities. Abplanalp asked Aldrich to please see to it that Rowland spent his time otherwise. Rowland, however, had firm support from the institution throughout the 15-year CFC-ozone ordeal. In his response to Abplanalp, Aldrich made clear his feelings that Rowland was a first-rate researcher, that the information he was producing was of extreme value, and that he would continue his work.

While Rowland also suffered some animosity from his own colleagues for becoming an advocate on a public-policy matter, he found defenders in the strangest places. One was Donald Davis of *Drug and Cosmetic Industry* magazine. Davis had a reputation in the aerosol business as an independent thinker who didn't always side with his industry. He had criticized industry officials for sometimes exaggerating their positions. While *Aerosol Age* magazine was an ardent supporter of industry, Davis' editorials in *Drug and Cosmetic Industry* were more objective. Davis, for example, described Rowland as a professional. "Rowland apparently has no ambitions toward becoming a Ralph Nader," he wrote. "An effective speaker always well prepared to support his arguments, Dr. Rowland has demonstrated a clear superiority of skills over his opponents from industry, over trade association officials and over those in the scientific fraternity who are not yet convinced."

Rowland didn't particularly enjoy the continual arguments. While it sometimes appeared as if he and Molina were delighted when they came up with additional evidence to support their theory, the fact was they weren't happy about it at all. The more they worked on the theory, the worse it looked for the ozone layer. They weren't trying to ruin an industry, only to protect the earth's environment. But Rowland wondered if some people thought he was crazy. "It looks and sounds like science fiction," he said of the theory. "I'm sure there were some who thought it was an elaborate joke on our part."

At a speaking engagement at one midwestern university, he warned students that science isn't always as dignified and clean as Americans would like to believe.

"When you're dealing with science on this level then everything is very polarized," he said. "And polarized science is not necessarily as much fun to take part in as some of the ordinary types of science I've been involved in."

But Rowland and Molina felt confident they could handle industry's arguments. Industry scientists, they felt, weren't as prepared to deal objectively and knowledgeably about stratospheric chemistry. "Industry scientists seem to be very far removed from the whole problem," Molina remarked at some gatherings of the two groups. "They have not been doing the right kind of work to be able to interpret what is happening with the fluorocarbons in the atmosphere."

The ACS meeting in Philadelphia reinforced this feeling and left both Rowland and Molina feeling satisfied. They had defended their theory effectively. Industry had little ammunition. Even reporters relatively unfamiliar with the issue described the meeting as a victory for scientists urging CFC regulations. "They had a field day," observed one.

As the conference wound up, Rowland remarked to Hal Johnston that the meeting had gone rather well.

"Don't get overconfident," Johnston responded. "They do not yet have anyone who knows very much about details of stratospheric chemistry. A year from now, they will have people who know something about this subject. And they will have found out how to muddy the water."

□ □ □

In 1975, *Playboy* suffered its first circulation drop. Automobile rebate programs became the latest gimmick to prod a sluggish economy. Miller Lite led an enthusiastic crusade for low-calorie beer. And sales of spray deodorants took a dive.

The chemical industry had two things to worry about in the fall of 1975: First, that the government might order a ban on CFCs, and, second, that consumers might not wait to be told what to do.

Industry was beginning to feel the effects of the Rowland-Molina hypothesis. And almost every segment of the CFC industry—from refrigerator salespeople to deodorant packagers—was sharing some blame for having failed to disprove this silly theory. They had gotten off to a poor start in late 1974 when a chemical industry trade association, the Manufacturing Chemists Association (which later changed its name to the Chemical Manufacturers Association), undertook of a study of CFC reactions in the troposphere, which, according to the

CFC-ozone depletion hypothesis, wasn't exactly the place to go looking.

Then the IMOS report had really sent things plummeting. The aerosols market was floundering. Precision Valve Corporation announced it had temporarily halted production at its Yonkers, New York, plant in order to consolidate manufacturing operations. The company cited unfavorable publicity, the economic recession, and high overhead as their reasons for shutting down the plant.

"The crucial question for the industry here and now seems to be whether or not consumer reaction will be ahead of government reaction," wrote Davis of *Drug and Cosmetic Industry* magazine.

Consumers had to decide for themselves if inadvertent modification of the stratosphere was for real or just another environmental scare. Some people did underestimate the American public's ability to understand the issue. The executive director of a trade group called the Chemical Specialties Manufacturers Association had especially grave doubts.

"If the chemistry intimidates us, think of what it must be doing to the consumer," Ralph Engel said at a meeting of aerosol packagers in October at New York's Americana Hotel. "I've heard that in a recent study, some people actually reported that they believed their underarm deodorant spray contains the harmful chemical ozone which, with regular use, can produce skin cancer."

Engel didn't cite the source of his study and there were no known reports in the mid-1970s of an increase in cancer of the underarm.

American consumers were wiser than they were sometimes given credit for. The CFC-ozone theory generated more letters to the federal government than any issue since the Vietnam War. Polls showed American consumers were well aware of the controversy. One survey showed that 73.5 percent of the respondents had heard something about it. Of those people, about half knew it involved the ozone layer. Almost half said they had stopped using aerosol products because of the threat.

The ozone issue did attract a bizarre segment. Novels featuring a world devoid of ozone became the science fiction rage. And environmentalists had the unwelcome support of several members of the Charles Manson gang, including murderess Sandra Good, who publicized their threats that injury or death awaited those who pollute the environment.

Shipments of aerosol cans were down by more than 25 percent in the first half of 1975, however, and it was clear consumers believed that the ozone layer might be in danger. With increasing skin cancer rates seen as a consequence of ozone depletion, there was little industry could do to defend itself without looking heartless.

Often, industry came across looking like "bad guys" while scientists on the side of the ban looked like national heroes. In May 1975, Sol Ganz, president of New York Bronze Powder Company, lashed out at the aerosol industry saying that its rebuttal against anti-aerosol publicity had been "totally ineffectual." Ganz had reason to be particularly upset. He had just received a disturbing letter which he shared with *Aerosol Age*.

The letter was from a young girl named Susan who said she and her friends feared getting skin cancer from ozone depletion.

"We don't want to die that way. My horse is going to have a foal this June & the foal will be alive in the year 2000 and I'll probably have to shoot her. All this will happen because of spray cans."

Susan and 23 friends signed the letter. Ganz wrote back and assured Susan that the destruction of the ozone layer was only a theory and that substitutes for destructive chemicals would become available if it proved correct.

Consumers weren't the only problem. Some industry insiders were shocked at how little loyalty was shown by their colleagues. The biggest blow to a united industry front came on June 18, 1975, when the Johnson Wax Company, the nation's fifth largest manufacturer of aerosol products, announced it would immediately end all uses of CFC propellants in aerosols. Samuel C. Johnson, chairman and chief executive officer, said at a news conference that the decision was made "in the interest of our customers during a period of uncertainty and scientific inquiry."

The company announced it would begin labeling products within 60 days to say: "Use with confidence, contains no Freon or other fluorocarbons claimed to harm the ozone layer."

The forces of state legislatures and the nation's youngest consumers had an effect on the Johnson Wax Company's decision to break ranks. According to Cicerone, a school child in Ann Arbor who was concerned about the CFC-ozone issue raging in her town had written to Samuel Johnson about his company's use of CFCs. In April 1975, Johnson replied in a letter to the child that "the fluorocarbon-ozone theory was crazy, similar to saying that the moon is made of blue cheese." Nevertheless, Johnson assured the child, his company was working on replacements for CFCs and may have found something. The child's letter was given to Cicerone who, in May, used it during

New York State Senate hearings on CFC regulations to counter claims by aerosol industry officials that no substitutes were possible for CFCs. Brandishing the Johnson letter, Cicerone said at least one company thought it could indeed get rid of CFCs.

More than 1,500 consumers wrote to the Johnson Wax Company in support of the decision. But Johnson's cohorts in the chemical industry were less than pleased. One industry official chastised Johnson Wax, telling a *Rolling Stone* reporter, "What they've done is to try to gain marketing advantage out of a difficult situation. I know damn well that's what it is."

While the aerosol industry tried to present a united front to fight CFC regulations, many aerosol packagers were resigned to eventual CFC regulations. As early as December 1974, some businesses that marketed hair-care products using CFCs were already working overtime to test nonaerosol versions of their products.

Coming up with substitute propellants wasn't easy. The known alternatives had problems. Vinyl chloride was a possible substitute, for example, but studies had revealed it could cause cancer. Isobutane and propane were very flammable. And nitrogen had a low boiling point. Nevertheless, manufacturers of the hydrocarbon and carbon dioxide propellants began an intense search for alternatives to CFCs. Other manufacturers addressed the idea of packaging products in nonaerosol dispensers, such as mechanical spray pumps, squeeze bottles and deodorant sticks. One industry vendor predicted in August 1975 that the market for spray pumps could double within the year. The packaging changes were so great by the fall of 1975 that an industry newsletter charged with listing new products gave up trying to keep track of them.

It was bad enough that some manufacturers were so quick to abandon CFCs, but what really peeved industry hard-liners were the companies that used the CFC industry's woes to promote their new products.

"Problem perspiration? Concerned about aerosols? Totally Honest non-aerosol anti-perspirant deodorant is the answer," read one advertisement for the new product. Advertising for the deodorant Ban Basic also contained copy claiming that the product "doesn't spray aerosol pollutants."

The prize for the most obnoxious commercial, claimed editors in *Aerosol Age*, was Mennen's "get on the stick" deodorant. The ad unfavorably compared an aerosol deodorant with the new stick type. "What is most amazing about this crude performance is that it is sponsored by an old-line, highly respected and conservative company which, incidentally, has been a leading marketer of aerosols," an editor in the journal *Soap/Cosmetics/Chemical Specialties* charged.

Efforts to fight CFC regulations while marketing alternative products produced a kind of schizophrenic atmosphere in the industry. One observer noted in *Aerosol Age* that, "The commercials pit different products of the same company against one another and seemingly work hard to destroy the consumer's faith in all products."

Those with the most to lose in the CFC-ozone debate fought the hardest. During the summer of 1975, industry had scored a somewhat effective, if very temporary, publicity coup by trotting out British scientist Richard Scorer. Scorer, of Britain's Imperial College, was a respected environmentalist, although one whose views "were widely divergent from other members of the 'save the Earth' club," noted one observer. Unlike other environmentalists, Scorer sided with industry on the safety of CFCs. A former editor of the *International Journal of Air Pollution*, Scorer was the author of several books on pollution and was very outspoken on what he called the unintelligent and dangerous use of science and technology.

Scorer was brash, articulate, and extremely suspicious of the CFC-ozone depletion theory.

"The only thing that has been accumulated so far is a number of theories," he said at one stop during his month-long U.S. tour. "People have gone mad over this business of monitoring things. They seem to imagine that if you monitor things enough, you'll find out how they work. You just might get confused."

The British scientist had no kind words for Molina and Rowland, calling them "doomsayers." Rowland responded to Scorer's attacks in his usual calm manner. "The gentleman is good at attacking. But he has never published any scientific papers on the subject."

Scorer's tour was organized by Hill & Knowlton, the world's largest public relations company, which was hired to help industry devise a strategy to counter negative publicity. Publicity wasn't always such a good thing, however. Hill & Knowlton booked Scorer on the TV talk show "The Firing Line" to debate Harvard's Mike McElroy. (Rowland turned down a request to debate Scorer.) Scorer was no match for McElroy's superb knowledge of the scientific aspects of ozone depletion.

Industry did better in the area of politics, often enlisting the most expensive lobbyists to fight state legislation on aerosol bans. And Du Pont spent millions of dollars on full-page newspaper advertisements, correctly assuming that no environmental or public interest group could afford similar measures.

Precision Valve turned to newspaper advertisements, too. In the fall of 1975 the company produced an emotional ad that claimed, "The consumer has been subjected to shock treatment by the very persons who keep emphasizing that the 'American people must be informed.' "

The ad noted that since the days of World War II, bug bomb aerosols "saved lives" and added that hair sprays would not have been possible without aerosol technology. The Precision ads took on a new note of desperation, however. "Don't give up on fluorocarbon aerosols YET!" the ad claimed. "We believe in what U.S. law holds clearly and we cherish dearly: you are innocent until proven guilty."

Whether chemicals had the same rights as humans was a concept that received an unfortunate amount of serious discussion throughout the CFC-ozone debate.

□   □   □

Whenever possible, industry tried to present a united front. At the November 1975 meeting of the Chemical Specialties Manufacturers Association at the Del Monte Lodge in Pebble Beach, industry representatives struggled to remain upbeat. Less than a hundred miles away, in San Francisco, Bay Area Air Pollution Control District officials held hearings on a proposal to ban CFCs. Threats were coming from every direction, and it was obvious that the CFC industry would have to stick together to fight this battle. One observer at the Pebble Beach meeting noted two industry VIPs talking together earnestly—not an earth-shattering event except that the two were old foes who hadn't spoken to each other for 32 years.

Despite its costly efforts, industry was losing in both the scientific and public relations arenas by late 1975. Industry writer Donald Davis sharply criticized aerosol businesses. He said that consumers were confused about which aerosols contained CFCs and which didn't.

"The real problem," Davis said, "is that no matter what industry says, consumers are buying fewer aerosols, and to say they are not will not make it so."

Du Pont officials also suspected that consumers were losing faith.

In a series of polls, consumers said they were using fewer aerosol products. Almost half of the respondents said the reason was danger to the environment. But among the study's most disturbing findings was the lack of faith in American business ethics. The Du Pont polls found that 29 percent of consumers questioned felt that marketers would put out an unsafe product if it were profitable. Another 35 percent thought "some" marketers would do so.

At times, aerosol manufacturers seemed downright desperate. In an enlightening summary of industry's defense to the problem published in an issue of *Drug and Cosmetic Industry*, one Du Pont official suggested putting product tags, shelf signs, and special displays around aerosol products saying "Feel Free to Buy Aerosols" or "Yes You Can Buy and Use Aerosols" to counter the adverse publicity.

Malcolm Jensen, president of the Can Manufacturers Institute, lamented that industry had been a "whipping boy" to Americans. He accused the Consumer Product Safety Commission of being the "newest monster created by Congress." Jensen also berated industry, saying it "has been too damned afraid to make a good case for itself. Too many in industry live in a sort of never-never land in which they think 'maybe they won't get me.' None of us wants to get our hands dirty. We don't take the initiative or the offensive. . . We have a good thing, let's fight like hell to keep it!"

□   □   □

Industry officials wanted proof. The Rowland-Molina hypothesis was based on calculations and limited knowledge about reactions between chemicals. There were no laboratory experiments that verified the theory. And, most significant in the eyes of industry, no one had taken measurements in the stratosphere to show that what Rowland and Molina said was happening was indeed happening. Until someone came up with the evidence, there should be no ban on CFCs.

Hal Johnston considered the argument a philosophical one. There was no such thing as absolute proof, he shot back. And how much proof did one require when the consequences meant destroying the ozone layer?

"You will never have absolute proof in this matter; science doesn't work that way," Johnston told a writer from *The Nation*.

Perhaps because of his experience with the SST debate and the CIAP studies, Johnston felt there was little to prove. He was considerably irked that the three-year, $50 million CIAP studies were practically being ignored. The Rowland-Molina theory, Johnston said, paralleled what happened with nitrogen oxides from natural sources and the exhaust from supersonic jets. "We have to change the labels in our equations," said Johnston to *Science News*. "But we don't have to redo all of the basic work." The chlorine theory, he said, is "proven by analogy."

The only chance that this was incorrect, Johnston said, was that some unknown aspect of chlorine chemistry would be discovered that inactivated or removed chlorine from the stratosphere.

There was also a rather rich philosophical argument over whose job it was to provide the necessary proof. Industry shunned the challenge of having to prove, after almost 45 years of use, that the chemicals were safe. This "negative burden of proof" approach has led "to overreaction and overregulation where there was not sufficient evidence demonstrating harmful effect," industry officials claimed.

Rowland and others felt strongly that it was the responsibility of industry to determine the consequences resulting from the manufacturing and use of its products. "The reason it looks like we have a new environmental problem every time we turn around is because we haven't looked carefully enough at the unfavorable consequences of all the chemical substances that have been introduced into the environment," Rowland told *Orange County Illustrated*.

However, it was always clear to Rowland, Molina, and others working on the problem that they needed to make measurements in the stratosphere.

Measurements, Rowland and Molina knew, would do two things: assess the accuracy of the computer models they were using and confirm what their models were saying.

There were three main elements to proving the CFC-ozone depletion hypothesis:

1. Evidence that CFCs were arriving in the stratosphere unaltered.

2. Evidence that CFCs in the stratosphere were broken up by the sun's ultraviolet radiation to produce chlorine atoms.

3. Evidence that the chlorine atoms initiated a catalytic reaction that destroyed ozone.

Shortly after taking their lumps at the ACS meeting in Philadelphia, industry representatives began assembling a scientific case against the hypothesis. Kaiser Aluminum & Chemical Company of Oakland, California, published a booklet entitled "At Issue: Fluorocarbons," which contained information from the company's top scientists. Kaiser summarized six scientific uncertainties with the Rowland-Molina hypothesis:

1. Measurements do not prove that CFCs rise to the stratosphere.

2. The proposed chain reaction had been based on lab experiments and had not been tested in the atmosphere.

3. The theory discounts other reactions in the proposed chain reaction that might nullify ozone depletion.

4. Much of the chlorine that can be found in the atmosphere is from natural sources, such as volcanic eruptions, and nature has handled this chlorine for thousands of years with no detrimental effects to the atmosphere.

5. Scientists have not yet shown the rate of significant reactions between substances in the atmosphere, that is, how easily certain substances would interact.

6. The theory does not take into account how chemicals would be mixed in air patterns in the atmosphere.

The fact that Mother Nature had handled massive amounts of natural chlorine for a millennium was a favorite argument of industry. For one, it was something the public could understand and, perhaps, find it in their hearts to believe. The planet is strong, resilient, and wouldn't allow us to be done in by a bunch of aerosol spray cans, industry officials exhorted. It was simply hard to imagine how the invisible gases could be so devastating. If the world was going to be destroyed by something, of course it would be war or aliens or a plague. To be wiped out by spray cans seemed so inappropriate, so, well, undignified.

"How can a world that has lived three decades with the specter of nuclear annihilation believe it will be done in by a blast of hair spray or underarm deodorant?" asked Michael Drosnin, a *New Times* writer.

Even Russell Peterson, the White House science adviser and a supporter of the Rowland-Molina theory, had trouble imagining it. "It is difficult to perceive that when you are spraying an antiperspirant in your bathroom you are endangering the health of the world," he said.

To some industry officials, one of the simplest measures of proof was to show that nature could cope with chlorine. And they just happened to have a belching volcano handy, poised to explode at any moment, that they believed would provide the proof they needed.

The volcano was located on Augustine Island in Cook Inlet, Alaska. The volcano study was being undertaken by the Council on Atmospheric Sciences, industry's scientific panel on the CFC-ozone issue. According to Dr. James P. Lodge, COAS science adviser (why a so-called scientific research group needed a science adviser was unclear), the volcano eruption would launch chlorine into the atmosphere. This natural phenomenon would confirm industry arguments that other chlorine atoms are propelled into the atmosphere and that the ozone layer is capable of sustaining such an attack.

Industry officials hoped fervently, while watching CFC profits sink in the summer of 1975, that this volcano would go off in a hurry.

Another somewhat unique approach to proving that nature had coped with chlorine in the past was developed by Reinhold Rasmussen at Washington State University. An experienced chemist who was funded by industry to study the Rowland-Molina hypothesis, Rasmussen had been measuring trace compounds in the air and went looking for old air trapped in corked bottles, antique hourglasses, and old water storage tanks. Rasmussen specifically wanted air before 1930 when CFCs were first manufactured. He would compare the old air with the present air to see if there was an accumulation of CFCs in the atmosphere now. The old air, he theorized, would show a high level of chlorine from natural sources.

Rasmussen was among the more diligent and creative of the scientists seeking to disprove the CFC-ozone depletion theory. He thought it was just another doomsday theory making scientists look silly. And he figured time would prove Rowland and Molina wrong. "Do you know the postmortems on most environmental 'threats' postulated in the '60s?" Rasmussen asked in *Science News*. "After the emotions were turned down, it was clear just how uncertain many of these really are."

Rasmussen worked hard to disprove the theory. Before measurements of CFCs in the stratosphere became so precise, he and atmospheric chemist William Zoller of the University of Maryland suggested that the chemicals might precipitate out of the atmosphere at the North and South Poles. With help from scientists at the Lawrence

Livermore Laboratory in Livermore, California, the researchers spec-ulated that the Antarctic could freeze out nearly all the CFCs produced each year. Zoller and Rasmussen based their theory on findings of unusually high concentrations of CFCs in ice samples from Antarctica. The finding was never explained. But the samples were stored in cold rooms which were refrigerated with CFCs, and the chemicals could have leaked into the air. (It was yet another irony of the CFC-ozone issue that the Antarctic was destined to play a major part in resolving the debate, but not at all in the way Rasmussen and the others had predicted.)

Even if they were unable to disprove the Rowland-Molina hy-pothesis, scientists (and not just industry scientists) continued to ques-tion just how severe the consequences of CFC emissions would be. Wouldn't the atmosphere find a way to cope?

"I'm personally convinced that the atmosphere of the planet has cycle and feedback mechanisms that are conducive to life on this planet," Du Pont executive Raymond McCarthy told *Newsday*. "I think it's conceivable, and this is theory, that nature has long since developed a cycle in which chlorine does not have deteriorating effects."

McCarthy could well have been paraphrasing the theories of James Lovelock who, in the 1970s, advanced his now famous and con-troversial Gaia hypothesis that the planet is a single organism with a series of checks and balances allowing for the existence of life. But even Lovelock, after initially scorning the severity of the CFC-ozone theory, later agreed that even a Gaian system could not resolve massive artificial fluctuations in the environment.

Other scientists argued that natural chlorine in the atmosphere was irrelevant. The influx of chlorine into the atmosphere from natural sources was part of a cycle that had evolved over millions of years. But the manufacturing of CFCs represented a sudden, rapid, and dra-matic increase of chlorine in the atmosphere—an increase that nature couldn't adapt to so easily.

There were other fallacies that had to be countered, most of which were undoubtedly dreamed up in public-relations offices.

Industry officials liked to claim, for example, that if the ozone depletion hypothesis were true, the amount of increased ultraviolet light would be no more harmful than what someone would encounter by moving from Duluth, Minnesota, to Biloxi, Mississippi. This argu-ment was illogical, other scientists pointed out, because the entire earth would be subject—involuntarily—to some increase in ultraviolet light, not just one person moving south. Plants and animals would also be subjected to the increase in ultraviolet light.

Another defense raised by industry concerned a mysterious increase in global ozone levels that were recorded in the 1960s. This statistic proved to be a difficult one for scientists to explain.

Like ripples over a pond, the ozone layer is amorphous and constantly shifting, becoming thin in some places and piling up in others. Since ozone is created by action of short-wavelength solar ultraviolet light playing upon oxygen molecules, the amount of shortwave ultraviolet light available to create new ozone is dependent upon a natural solar cycle that in some years produces more ultraviolet light than during other years. Researchers have found that increases in ozone seem to occur during peak periods of activity, such as sunspots, on the surface of the sun. This activity occurs in 11-year cycles.

Scientists had begun taking measurements on ozone since the 1930s and had noted a 7.5 percent increase in ozone during the 1960s. To critics of the ozone depletion hypothesis, this was proof that the ozone layer was constantly shifting and that losses couldn't be pegged on human activity. But there were some flaws in that argument. For example, because CFCs take from 50 to 100 years to build up in the stratosphere, measurements of ozone taken in the 1960s would not have detected the great influx of the chemicals that began in the 1950s. The measurements also failed to take into account how much greater ozone would have been if not for CFCs. Natural increases in ozone from sunspot activity could easily disguise ozone reduction from CFCs by offsetting it. But when the sunspot activity declined again to produce less ozone, the decrease from CFCs would be more evident than ever. The difficulty in reconciling artificial ozone loss with the natural sunspot cycle continues to plague researchers today who are charged with measuring total global ozone.

Rowland recognized that there was only one way to settle the matter and convince the U.S. government and the National Academy of Sciences that CFCs were a serious threat to the ozone layer. They would need evidence to show the CFC-ozone reaction in the stratosphere.

The first step, detecting the chemicals in the atmosphere, was fairly easy. Throughout 1975, researchers used an arsenal of high-flying aircraft and balloon-borne measuring devices to detect CFCs 15 miles or more above the earth's surface. In July 1975, researchers from the National Oceanic and Atmospheric Administration and the National Center for Atmospheric Research (NCAR) in Boulder, Colorado, showed for the first time that the chemicals reached the stratosphere, confirming that there were no major sinks for the CFCs in the troposphere. The two groups' studies also indicated that the chemicals were being broken up by ultraviolet light because measurements showed the amount of CFCs present in the stratosphere dropped off with altitude. These measurements were, in fact, so close to those predicted by computer models that one member of the research team called the findings "absolutely astonishing."

"In our opinion, these results demonstrate directly that fluorocarbons are transported into the stratosphere with no large loss in the lower atmosphere," said NOAA scientist Art Schmeltekopf. "These direct observations also suggest that it is unlikely we will find major tropospheric sinks for these compounds."

The second of the three steps needed to prove the hypothesis was taken only a few months later when several groups of scientists ·confirmed that CFCs dissociated by ultraviolet light did give off chlorine. The credit for that discovery, at least in the general media, went to scientists Pierre J. Ausloos and Dr. Richard E. Rebbert of the National Bureau of Standards who reported their evidence at an August meeting of the American Chemical Society. While other scientists weren't all that impressed with the NBS findings—after all, other people had discovered the same thing—the study was clearly the big news of the meeting among reporters. A *Rolling Stone* reporter (the CFC-ozone issue was controversial enough to attract all kinds of media) noted that even impartial scientists weren't immune to a little publicity. The NBS, a latecomer to the ozone depletion issue, announced its findings in advance of the ACS meeting by mailing letters, a five-page press release, and a copy of its report to selected scientists.

"They were always announcing with a press release what everyone else had accepted for six months," Rowland chuckled. Regardless of who got credit, the new evidence was also reported in September in hearings before the Senate Subcommittee on the Upper Atmosphere and received widespread attention. Now industry's only remaining argument was that there was no proof that the chlorine initiated a chain reaction that would destroy huge amounts of ozone. A Du Pont representative called this "The most speculative and most critical aspect of the theory."

To prove the existence of a catalytic reaction, scientists had to look for other substances given off during the chain reaction. In particular, scientists were looking for a substance called chlorine oxide, which was the product of a reaction between chlorine atoms and ozone. This was important. In the fall of 1974, Rowland had sat at his desk and scribbled a list of a dozen things people could do to confirm the theory. The number-one item on the list was the detection of chlorine oxide in the stratosphere.

To scientists, chlorine oxide became the "smoking gun" of "ozonegate." Finding this chemical would convince the world that CFCs were destroying the ozone layer. If they were unable to detect this chemical, it might signal that the theory was wrong and the ozone layer was safe after all. By the fall of 1975, research projects on this third step were underway at NOAA, NBS, the Department of Defense, EPA, the Energy Research and Development Agency, and NASA.

Researchers finally did locate their smoking gun, but not before a new piece of evidence surfaced that nearly shot the entire CFC-ozone hypothesis to bits. Ironically, it wasn't industry that threw the theory into question. This evidence surfaced in front of Mario Molina's disbelieving eyes on the Irvine campus during the spring of 1976.

5
Aerosol Ban
January 1976—May 1977

Sherry Rowland was growing impatient. It was January 1976. It had been two years since he and Molina had made their discovery. He felt that the evidence gathered over the past year was enough to convince the nation's policy makers that a ban on the nonessential uses of CFCs—not to mention restrictions on other uses of the chemicals—was in order.

The ban must come now, Rowland told Lydia Dotto of the *Toronto Globe and Mail*. The less well understood details of the chemical reaction "aren't going to be well known for 10 to 20 years. We're going to have to make whatever decisions we make without knowing whether there's a potential for climatic catastrophe."

Others agreed with Rowland. At the 12th International Symposium on Free Radicals held in Laguna Beach, California, in January, Paul Crutzen reported evidence that photochemical events caused abrupt and significant decreases in stratospheric ozone. Crutzen, a short man who wore glasses, was born in the Netherlands and had done much of his work on the atmosphere while in Europe. A cheerful person who loved to hike and skate, Crutzen had left Sweden the previous year to join NOAA and the National Center for Atmospheric Research in Boulder. Perhaps because he was not a U.S. citizen, Crutzen was removed from the political battle over CFCs. Years later, however, after returning to Europe to direct the Max Planck Institute for Chemistry in Germany, Crutzen became involved in the debate over the effects of nuclear winter—the climatic changes that could come about through the use of nuclear weapons.

In the early 1970s, Crutzen had made a very specific prediction. He suggested that ozone would be lost at high altitudes and at high latitudes. In order to prove his theory, he asked NASA to look at satellite records for ozone in the north polar high stratosphere following a large solar flare that had taken place years earlier. Calling the evidence "the experiment that nature did for us," Crutzen's examination of the solar flare—the largest ever reported—showed it had produced enough nitric oxide to confirm his prediction.

Nitric oxide, of course, was the substance that so alarmed scientists during the SST debates because the chemical was present in the jets' exhaust. In January, however, the World Meteorological Organization announced that, based on new, more accurate calculations, SSTs did not present as great a threat to the ozone layer as originally estimated. The jets were, eventually, given limited landing rights in the United States based on studies which showed that depletion of ozone was directly dependent upon the altitude of the flight of the aircraft. With planes flying at altitudes below 12 miles, negligible changes in ozone levels were expected. (Like the ozone depletion issue, opinions on the SST issue continued to vacillate from "severe" to "no problem." Today, it is a serious concern.)

The World Meteorological Organization also stated that new calculations showed that the CFC threat to ozone was greater than originally anticipated. But the waffling on the environmental hazards from the aircraft incensed old SST proponents and chemical industry officials. See what happens, they said, when the government issues regulations before all the scientific evidence is in?

They had a point. Science is hardly infallible. History is full of tales of scientists who made great discoveries one day only to have their work disproved years later. Even some Nobel prizes have been awarded for work that later turned out to be false. Although they would hate to have it happen to them, most scientists feel such that a twist of fate is nothing to be ashamed of. After all, science would not progress without men and women who are willing to postulate theories and see them disproved in favor of better, more complete theories. When it comes to environmental matters, many feel that it is better to err on the side of environmental safety than to risk causing irreparable harm to that which sustains life on earth. That was Rowland and Molina's position all along. And it is to their credit that if anyone was going to find the fatal flaw in their theory it was going to be them.

☐ ☐ ☐

To help prove their theory was correct, Rowland and Molina had to know if there were other chemical reactions or processes taking place in the atmosphere that might offset the effects of the CFC-ozone reaction.

Molina, who was happiest pouring over obscure journals in the library or fiddling in the laboratory, had for weeks been making a systematic check on the substances that could figure in potential reactions. In particular, Rowland, Molina, and chemist John Spencer, who had joined Rowland's team at Irvine, were interested in some compounds that were expected to have a very brief life and, therefore, probably wouldn't figure into ozone depletion. But they had to be sure. In what he later credited to his year of study at the University of Freiburg, Molina decided to explore some old German chemical literature and came across a molecule called chlorine nitrate.

In the summer of 1974, Rowland and Molina had considered chlorine nitrate but thought that it was a short-lived reaction, and, therefore wouldn't have much effect in the atmosphere. They had expected that the lifetime of chlorine nitrate would be only minutes, but experiments showed the substance was more stable and might stay intact for hours. This finding threw a huge new loop into the theory. The presence of chlorine nitrate, Rowland and Molina knew, would make a difference in the chain reaction. Chlorine oxide, formed when chlorine is released from fluorocarbons, might link with nitrogen oxides in the atmosphere to produce chlorine nitrate. By tying up chlorine, less ozone would be destroyed than the 7 to 14 percent Rowland and Molina had suggested. This could indeed be the sink that Rowland and Molina had long ago looked for and dismissed as nonexistent.

Just how much this changed the calculations for ozone loss was now up in the air. With a sinking feeling, Rowland dutifully put in a call to the National Academy of Sciences, whose report was due out in a matter of weeks. Julius Chang, a mathematician with the Lawrence Livermore Laboratory, was doing the model for the NAS panel and quickly inserted the new information into his computer to see how much it would change ozone depletion estimates. Chang was stunned. He picked up the phone and called George Pimentel at Berkeley, who was also on the NAS committee.

"It reduces the calculations by a factor of three," he told Pimentel solemnly. He paused before adding the clincher, "And it changed the sign."

Instead of having 14 percent less ozone, it appeared from Chang's calculations that chlorine nitrate might actually *increase* ozone by 5 percent.

"It can't be right," Chang said.

Chang suspected, as did Rowland and Molina, that while chlorine nitrate might reduce ozone depletion slightly, there were also flaws in the models that were making the chlorine nitrate out to be more influential than it probably was. For one thing, the calculations implied that there was a great deal of chlorine nitrate in the lower atmosphere, but direct measurements didn't support that.

The news that the CFC-ozone depletion hypothesis was in question was circulating within days, and everyone who had a computer model began figuring in new calculations for chlorine nitrate. The estimates ranged from slightly reduced ozone loss, which Crutzen's model showed, to an increase of ozone. Rowland and Molina estimated that ozone depletion would be only half as bad as they had originally suggested. They also pointed out that that was still a hazardous amount of depletion. But others said 90 percent less ozone would be destroyed, a figure that would practically exonerate CFCs. To confuse the matter even more, there was a suggestion that hydrogen chloride, which was known to be a stratospheric sink, might react with the chlorine nitrate, canceling out both sinks. In that case, ozone depletion would be even worse than originally estimated.

In the lab, Spencer and Molina learned how to handle the chlorine nitrate, which was no easy feat since the substance is hard to prepare, and air mailed a sample to Doug Davis, a chemist at the University of Maryland. Davis' calculations showed that the reaction of chlorine nitrate and oxygen was very slow, and that this wasn't the main source of removal of chlorine nitrate. Davis found that the depletion would be one-half to two-thirds the predicted rate but insisted the figures were preliminary. And they were later disproved. But much to Rowland and Molina's dismay, industry immediately latched onto a Du Pont scientist's calculation showing reduced levels of depletion and used it to try to convince the American public that the CFC-ozone depletion theory had been disproved.

For Rowland and Molina, the next few months were distressing. It was bad enough that the chlorine nitrate finding had generated so much confusion, and it was all done so publicly. But the Irvine chemists' distress was nothing compared to the aura of grief that permeated the NAS panel. With a rough draft of their report written and nearing finalization, the panel now had to reconvene and figure out what chlorine nitrate really meant.

"I went through a nervous disorder with chlorine nitrate," panelist Fred Kaufman, a chemist at the University of Pittsburgh, admitted to his colleagues at a scientific conference later that year. "It was a terrible thing. We had a document finished which had results which were even larger . . . 13 percent possible long-term depletion, and then chlorine nitrate appeared."

It wasn't that Kaufman, or anyone on the committee, faulted Rowland or other scientists trying to figure out the mysteries of the stratospheric chemistry. Indeed, it was typical of the scientific process that new information would be found and theories revised. But, Kaufman noted, most science isn't conducted with government, industry, and environmentalists breathing down your neck. "What's holding you up? When are you guys going to come up with your report?" panelists heard throughout the spring and summer.

The panelists were weary. The committee work had been exhausting and fraught with logistical and personality problems. Members of the panel were sometimes required to spend two weekends a month in Washington. And they depended on the cooperation of all the scientists involved to help them get the report out. But the various groups doing the computer models were competitive, and it wasn't always easy to bring them together to give the panel a cohesive report on estimates of ozone depletion. When the chlorine nitrate bombshell dropped, tempers flared. "It was back to square one," Canada's Harold Schiff, a committee member, said.

The news was, however, just what the beleaguered chemical industry had been looking for.

In the spring of 1976 headlines across the country suggested that the CFC-ozone depletion theory was wrong, that the ozone layer was in no danger, and that spray deodorants were safe.

"Rumor and Confusion Follow Ozone Theory Revision," said one headline. And, "Aerosol Scare May Be Over," claimed another. Precision Valve pounced on the chlorine nitrate dilemma. In an ad targeted toward aerosol packagers, the company reprinted a news item quoting Rowland that the new estimates showed depletion may not be as severe as once thought. Entitled, "Who's Out on a Limb?" the ad stated that "important new evidence does not support the ozone-depletion theory." It concluded, "Keep the faith. Consumers prefer aerosols."

On May 12, industry's Council on Atmospheric Sciences held a press conference in New York's Barclay Hotel to say that earlier reports of the ozone layer's demise had been greatly exaggerated. Dr. James Lodge of the COAS used the occasion to ask for "a fair trial, not a lynching party." Scientists, however, were still grappling with the chlorine nitrate findings and the issue hadn't been resolved. Groups like New York's Public Interest Research Group weren't ready to exonerate CFCs and demonstrated outside the Barclay, passing out "ban the can" pamphlets.

"How in the hell can they do that?" one industry official was heard muttering. "Right in front of the door!"

But things had certainly improved rather suddenly for industry. At the Chemical Specialties Manufacturers Association mid-year meeting in Chicago, the mood was decidedly upbeat. "Thank God we didn't panic altogether two years ago," Abplanalp told the *Detroit Free Press*. "Business is back to damned near normal." To British scientist James Lovelock, the whole business was beginning to make a little more sense. The chlorine nitrate finding "takes the heat out of the situation," he told the *Observer*. "We are at least now back in the realm of normal science."

The chlorine nitrate confusion seemed to be just what industry needed. Not only did the new calculations appear to disprove the Rowland-Molina hypothesis, the evidence had come from Rowland and Molina themselves.

Only the extent of ozone depletion was in doubt, however, and the Irvine researchers stood by their call for the ban. Rowland warned industry not to depend on the chlorine nitrate reaction to clear things up. "The bottom could fall out from under them very soon if they persist," he told *Food and Drug Packaging* magazine.

The bottom did fall out. At the insistence of the NAS committee, a semi-emergency meeting of the modelers was called one weekend at NCAR in Boulder. Crutzen and Cicerone both agreed to meet in Boulder and run their computer models all weekend, if necessary, to try to come up with a reasonable estimate of just how much chlorine nitrate changed long-term estimates of ozone depletion. It turned out to be a wild weekend.

By late Sunday night, the modelers were close to resolving the enormously complicated problem. At 3:00 A.M. on the last day, the scientists met to review their results. All had found that chlorine nitrate caused the models to compute very small total losses of ozone. Cicerone, however, noticed that the reason the models were showing little ozone change was because huge losses of ozone in the upper atmosphere were being compensated for by large ozone gains in the lower atmosphere. This altitude profile of ozone for a planet that was being loaded with continued CFC emissions was unlike any that had ever been measured before in the real atmosphere. Studying Cicerone's graph showing the weird shapes, Crutzen remarked, "This is not my planet," and began to fret that climatic effects of ozone redistribution might turn out to be more dangerous than the effects of ozone loss.

Cicerone and Crutzen were pressured by NAS president Philip Handler to fly to Washington as soon as possible to inform the panel of their findings. Arriving at the academy the next day, Cicerone and Crutzen were met by an anxious Handler. A stately looking person who was an eloquent speaker and adept at handling sticky political issues, Handler pulled the scientists into a private office and sought their assurance that the report nearly all of Washington was awaiting wouldn't be significantly delayed by this chlorine nitrate mess. Cicerone sympathized with Handler's problem. The press knew the report was due, and any delays would set off suspicion that the theory was flaky or that the academy was caving into industry pressures and holding back information from the public. Handler glanced at the two scientists nervously, waiting for one to speak. Since Crutzen was the veteran of the two, Cicerone thought Crutzen should explain the matter. Crutzen stared at Handler blankly.

"When we ran the model calculations it looked like this was no longer the atmosphere of the earth," Crutzen stated slowly as Handler paled. "When I looked at the graph I said, 'This is not my planet.' "

Intelligent person that he was, Handler's first decision was that Crutzen and Cicerone should not be allowed to relay this most upsetting conclusion to the press. But, after explaining their findings in more detail, Cicerone and Crutzen convinced both Handler and the academy that the problem was not that impossible. Revisions to the model showed that chlorine nitrate would have the effect of changing the distribution of ozone in the atmosphere with more ozone low in the atmosphere and less up high. Crutzen and Cicerone persuaded the academy to continue with the report.

And, after continued weeks of haggling and efforts to improve their computer models, scientists agreed that the effect of chlorine nitrate appeared minimal. While the substance did reduce the amount of ozone depletion originally suggested, there were errors in the models that had made the original estimations of increased ozone completely false.

In the end, most models agreed that the amount of ozone depleted would probably be on the low end of the 7 to 14 percent Rowland and Molina had originally suggested—still a severe problem.

It was hard to tell who felt worse at this point. The scientists knew that the chlorine nitrate controversy had thrown the NAS report into a state of confusion. Meanwhile, industry had just "blown" its one chance to put an end to the entire controversy. The June 1976 issue of *Aerosol Age* contained another post-mortem report, a brief news item entitled "Scratch One Volcano." The magazine reported that the long-awaited volcano had fizzled. COAS science adviser James Lodge explained that too few stations were in place to monitor the explosion because it had gone off too fast. But, noted *Aerosol Age* editors, it was hard to believe that after nine months of waiting they were unprepared to monitor the thing. "Betting on that volcano didn't help," the magazine noted. "The industry blew that one (pun intended)."

Rowland, Molina, and their colleagues felt badly, too. They were depressed by the criticism and by the timing of the chlorine nitrate flap. The past few months had been brutal. For the first time since he had embarked on defending his theory, Molina felt he had been put down by an industry official at a scientific conference. Trying to defend the theory by suggesting that the chlorine nitrate finding would have no great effect, Molina was interrupted by an industry chemist. "Look," the man said, "the whole thing looks like nothing. We can even get enhancement of the ozone with this chemical!" Lacking the data to argue otherwise, Molina was forced into silence.

Rowland also seemed subdued. That spring he told a home town reporter that he wanted to get on to other research projects, his students, and his tennis game. "It would be nice if it would get settled," he said. "I get impatient sometimes. But I've never considered walking away from it. When you work in the fundamental sciences, one of the questions you keep asking yourself is, how valuable is my work? Are the answers I'm getting or hope to get important? I haven't needed to ask myself that in quite awhile."

Although the NAS Panel on Atmospheric Chemistry had the revised figures on the chlorine nitrate effect in hand and could see that
the theory hadn't really changed, the controversy had cast doubt on
the scientific process in general. An editorial in the *New York Daily
News* summarized the doubt this way: "Now that the scientists have
been put in the position of crying wolf, who will listen to new warnings?"

Despite the embarrassment and tension the chlorine nitrate findings had caused, Rowland and Molina were applauded by many scientists for being the ones to bring the problem to attention. After all,
Cicerone said when a worried Rowland had called him to tell him
about the chlorine nitrate findings, "one of the hardest things to do
in life is be self-critical, to find our own mistakes and to think harder
than we thought before."

In Washington, policy makers who had been active in pushing
for legislation to control CFCs were stunned by the chlorine nitrate
episode. In touch with scientists on a weekly basis, Moomaw (the
scientist working for the environmental subcommittee) advised Bumpers to remain silent while scientists worked the issue over. To many
people, legislators and scientists alike, however, the very fact that it
had been Rowland and Molina who had come forward with the new
information somehow enhanced the credibility of the entire issue.
While some critics had contended that Rowland had made up his mind
that the CFC-ozone theory was correct, it was now apparent that he
was still working, along with others, to learn more about stratospheric
chemistry and that he would contribute to the downswings as well as
the upswings in depletion estimates.

And, despite the unfortunate timing of the incident—with the
NAS on the brink of making its final recommendations on CFCs—
many felt the science community handled the problem professionally.

"People behaved like scientists," Cicerone later recalled. "They
talked about all the possibilities."

The chlorine nitrate controversy delayed the NAS report for five
months. The report was revealed on September 13, 1976, in Washington at one of the most heavily attended press conferences the NAS
had seen in a long time. There were two reports. One, a study by the
Panel on Atmospheric Chemistry, issued a verdict on the validity of

the Rowland-Molina hypothesis. The other, written by the Committee on Impacts of Stratospheric Change, made recommendations to the federal government on how to proceed on proposed CFC regulations. Perhaps if the NAS had stuck to merely validating the scientific hypothesis it would have been better off because the report on recommendations confused people more than ever.

The report concluded that CFCs were damaging the atmosphere and should be restricted. It estimated that CFCs released at 1973 rates would eventually produce an ozone loss of 2 to 20 percent with 7 percent being the most likely. Rowland and Molina had been correct. The report also agreed with the theory that decreased ozone would lead to an increase in the amount of ultraviolet light reaching the earth and, therefore, could be expected to cause more skin cancer. The release of CFCs, the report continued, could also initiate climatic changes and contribute to the eventual warming of the earth's atmosphere by the "greenhouse effect." But, the other NAS committee advised, government should be given two more years to gather evidence and assess the severity of the problem before issuing regulations on CFCs.

The feeling in Washington was that the committee had bowed under its burden of having to decide on the economic fate of one of the nation's biggest industries.

"We were all conscious of the fact that (regulations) would be an outcome," said panelist Schiff of the report. "It made us a bit more cautious, perhaps, in phrasing and drawing conclusions because I think we were responsible enough to realize we were fooling with a fairly major industry."

Legislators who had been waiting for the report before deciding to act on amendments to the Clean Air Act were also less than impressed with the academy's bottom line. For months, legislators and their staffs had become aware that the NAS panel members seemed exceedingly uncomfortable with their role. Milling around the academy headquarters after the announcement, Moomaw sensed disappointment in the air. The academy had waffled and everyone knew it. Nevertheless, Bumpers and others decided that the report was conclusive enough to push ahead with amendments to the Clean Air Act.

Rowland awaited the news of the NAS report with some trepidation, setting up tape recorders to monitor radio reports. It was noon in Washington, 9:00 A.M. in California. Rowland spent the better part of his morning trying to deduce what the report had said.

"The first announcements were like the report—sort of mixed,"

Rowland recalled. "It was like saying 'Yes, but . . .'. You could hear some of the but and some of the yes. It wasn't until we heard two or three announcements that it said scientifically that, yes, chlorine was going to attack the ozone. And then, however, that we do not think we should do anything about regulations."

The press was equally confused as to what the report was actually advising. "Scientists Back New Aerosol Curbs to Protect Ozone in Atmosphere," read a *New York Times* headline the next day, while the *Washington Post*'s account was entitled, "Aerosol Ban Opposed by Science Unit."

A week after the report was released, at a scientific conference on ozone depletion at Utah State University, the panel's Kaufman expressed his regret at the "uncertainties" contained in the report. Addressing R. David Pittle, commissioner of the U.S. Consumer Product Safety Commission, Kaufman expressed his fear that the members of the panel would be tarred and feathered and asked the commissioner if he knew of any noncarcinogenic tar.

To Rowland and Molina, however, the news was vindication. (They clung to the 1976 NAS decision in later years when doubts about their theory surfaced again). If Rowland was a little disappointed in the panel's lack of a stand on a ban, he was also understanding. "You could see where they might feel a little bit nervous about the ease with which it went from 14 to 7 percent," he said. "It was appropriate scientific caution coming out." The Irvine chemists felt that the report suggested CFC use should be curtailed immediately. Yes, the uncertainties in the amount of ozone depletion were large, but that didn't change what they considered to be the prudent course of action.

Industry representatives, meanwhile, said the report confirmed that more research was needed before CFCs could be regulated. In an advertisement appearing in newspapers around the country the following week, the Council on Atmospheric Sciences quoted the NAS report in bold type. "'We wish to recommend against a decision to regulate at this time.' We agree!"

Privately, however, the men and women who made, packaged, marketed, and sold CFCs and products containing CFCs were as confused by the report as anyone. Their reaction became apparent to Rowland at the International Conference on Problems Related to the Stratosphere at Utah State University in Logan a few days after the report was released.

Throughout the three-day meeting, Rowland had heard rumors that the industry people were unhappy about the report. The committee had been wishy-washy. Saying that CFCs were a problem but then refusing to do anything about it was a cop-out, many felt. To Rowland, however, there seemed to be a strange aura of acceptance, or perhaps resignation, among the industry representatives at the meeting. Toward the end of the conference, attendees were invited to a picnic in a woodsy area in the hills above Logan. It had been a nice September day and the group had played volleyball and snacked until dusk. There had been little shop talk at the outing; but now, in the semidarkness, Rowland suddenly found himself face to face with some of the people who had vehemently opposed him and his theory for almost two years. One by one, they sauntered up, shook his hand, and hastily congratulated him. Rowland strained to see in the darkness but couldn't read their name tags. And he got the distinct but eerie feeling that they wanted it that way.

"It was an opportunity for some of them to say, I guess, that they thought the science was interesting and impressive or something of that sort but without having to be held accountable for saying it," Rowland later recalled.

Two things were certain when the sun rose on picturesque Logan in the morning, however. The earth's ozone shield was still being attacked by chlorine. And the chemical industry had only just begun to fight.

□ □ □

Industry's claims of victory following the National Academy of Sciences report were short lived. While the report had left many Americans perplexed over the fate of their spray-can deodorants and hair sprays, at least a few government officials were determined to wrest a decision regarding CFC regulations from the report—no matter how cleverly the panelists had worded the document to avoid doing just that. The first clue that a ban on uses of CFCs in spray cans was imminent was dropped during a speech by Russell Peterson at the Logan conference.

"It seems to me, the basic concern of this conference is doubt—scientific uncertainty and the proper course of action when one is confronted by it," Peterson said.

For the NAS panel, the response to doubts over how much chlorofluorocarbons would deplete the earth's ozone shield was to recommend two more years of additional research before attempting to regulate use of the chemicals. To Peterson, this hesitancy suggested that the benefits of using aerosol sprays outweighed future environmental catastrophe. In his speech, and later, before a packed press conference in Logan, Peterson recommended that government officials move forward with restrictions on nonessential uses of CFCs and suggested that consumers stop using products containing the chemicals. The makers of a chemical should be required to demonstrate that their product is not responsible for causing harm, he said.

"The party that stands to benefit from the introduction of a new activity or substance should bear the cost of showing that claims of damage . . . are outweighed by its benefits," Peterson said. "I believe firmly that we cannot afford to give chemicals the same constitutional rights that we enjoy under the law. Chemicals are not innocent until proven guilty."

But Peterson did not absolve scientists from addressing the many remaining questions about ozone depletion.

"From the pure scientific perspective, there remain valid doubts about the effect of fluorocarbons on the ozone shield. From the public-policy standpoint, however, there remains no valid reason to postpone the start of the regulatory procedures."

Peterson was not alone in his decision to call for a ban. As he took his place at the podium in Logan, R. David Pittle, Commissioner of the U.S. Consumer Product Safety Commission, cited the academy's findings that increased ultraviolet light and climatic changes would accompany ozone loss.

"The presence of additional supporting evidence and findings indicating an increased magnitude of risk lead me to seriously question a recommendation for an additional two years' research. . . I particularly question this delay as it seems highly unlikely that the additional studies will result in a recommendation that the use of chlorofluoromethanes as aerosol propellants not be banned."

And, in the third strike to industry, Wilson K. Talley, assistant administrator for research and development at the Environmental Protection Agency told the conference participants in Logan, "We just cannot postpone decisions on these problems indefinitely in the hope that better data may be available in the indefinite future."

Requiring "body counts" before dictating public policy was unacceptable, Talley charged.

"I am convinced that the public interest demands precautionary environmental regulations, based on the best data available, early enough to assure that no such 'body counts' are ever needed."

Within days after the conference, the creaky wheels of the regulatory process were in motion. The 1975 report from the committee on Inadvertent Modification of the Stratosphere had instructed government agencies to begin issuing regulations on CFCs should the NAS committee find evidence that the Rowland-Molina hypothesis was correct. On October 15, one month after the release of the academy report, the U.S. Food and Drug Administration proposed a phase out of all nonessential uses of CFCs in food, drug, and cosmetic products. The EPA followed that same week by announcing that it, too, was seeking a ban on nonessential CFC uses.

The news rocked the CFC industry, which had had barely enough time to savor the reprieve the NAS had given it before being sentenced to death by a new jury. Du Pont officials were outraged.

"We find the FDA's intent to establish a timetable for phasing out fluorocarbon aerosols astonishing," the statement said. "When NAS wrote in its report 'we recommend against the decision to regulate at this time,' the academy clearly recommended against the kind of action that FDA is planning."

But it was too late for industry to muster the kind of quick counterattack it needed to prevent a ban of nonessential CFC uses. On May 11, 1977, the EPA, FDA, and CPSC jointly announced a timetable for the phase out. Manufacturing of CFCs for nonessential products would cease by October 15, 1978. Companies would be required to stop using CFCs in nonessential products by December 15, 1978, and interstate shipments would be banned by April 15, 1979.

The regulations generally affected the spray-can industry. In defining products that were considered essential, and thus exempt from the ban, the government stipulated that a product must provide a public benefit that could not be obtained without the use of CFCs, which was certainly not true of cans containing CFC refrigerants used for chilling martini glasses. "Essential" also meant that the product did not release significant amounts of CFCs, or if so, that the use was warranted. Finally, products were exempt when there was no practical alternative available. The government's determination of products that could be considered 'nonessential' rankled industry further. The marketplace, not government, should make such designations, industry officials charged.

While there were few ready alternatives for the use of CFCs in refrigerators, air conditioners, and cleaning solvents, aerosol producers had had two years to anticipate a ban of the chemicals and spray cans and they responded with the kind of entrepreneurial spirit and technological skills that make America great. By the spring of 1977 it was hard to believe that some of the people who had described the CFC-ozone depletion theory as ridiculous were now happily touting the latest spray pumps and non-CFC propellants. Suddenly, it seemed as if industry held a certain pride—if not a distinct marketing advantage—in promoting products that were safe for the environment. Arrid Extra Dry deodorants rolled off the assembly line with the slogan "Safe for the Ozone" stamped on the can.

Economically, the aerosol industry was to suffer little from the 1977 ban. The very day after the federal agencies announced the phase out, Robert Abplanalp unveiled his latest invention, an idea that he said he had come up with during six months of doodling on sketch pads. While still insisting that the CFC-ozone depletion theory was a hoax, Abplanalp said his new invention would have "wiped out fluorocarbons anyway."

The device was a new valve, given the trade name Aquasol, that operated on a mixture of water and butane gas as a propellant. The product used no CFCs, and because the butane floated in pressurized form at the top of the can, more of the can's space could be filled with hair spray, deodorant, or whatever. Abplanalp had, with one swift stroke, dumped CFCs and quieted consumer advocates who claimed that spray cans weren't cost-effective since they contained too much propellant and too little active ingredient. Aquasol, Abplanalp noted, was nonflammable, and unlike spray pumps, delivered a fine, even mist.

The Aquasol announcement seemed to assure Americans that they could have their ozone and hair sprays, too. *Time* magazine announced the Aquasol invention on May 23, 1977, in a story that featured a picture of Abplanalp and his new spray. The picture bore the caption: "A potential savior." The chagrined environmentalists who had fought bitterly with Abplanalp on the CFC-ozone issue for so many months could only assume that "savior" referred to the new valve.

While Aquasol was touted as the savior of the industry, much of the aerosol market simply switched to a hydrocarbon propellant that had been in use for several years.

The future looked bright that spring for an eventual total ban of all CFC uses. In their book, *The Ozone War*, Dotto and Schiff pointed out that annual spray-can sales had fallen from a 1973 peak of 3 billion to less than 2.3 billion, the 1968 level. "It is clear that a process has begun that will almost certainly result, in the United States, in strict controls of all releases of fluorocarbons to the atmosphere, if not necessarily a total elimination of their use," Schiff and Dotto wrote.

And, in announcing the phase out of nonessential CFCs, the EPA said that it would eventually allow the chemicals to be used only when the manufacturer could demonstrate that they were essential for safety and the effectiveness of the product.

But the aerosol ban was only one battle in the ozone war. And at least a few people realized that.

In the offices of the Natural Resources Defense Council, which had, in December 1975, filed a second petition calling for regulations on aerosols containing CFCs, the announcement of the ban was met with joy. But it was not an occasion to break out champagne. This was a no-nonsense, if somewhat jaded, group.

"We have enough humbling experiences here that when we do win a battle we know that the battle is never over," Ahmed later recalled, soberly. "We knew we had won only one part of this battle."

6
The Dark Years
June 1977—October 1980

For an environmental movement that had started out so convincingly, the next eight years proved astonishing as the momentum to protect the earth's ozone layer from manmade pollutants crumbled. From 1977 to 1985, Rowland, Molina, Cicerone, and a small group of environmentalists devoted to ozone protection watched in near disbelief as the burden of proving that CFCs destroyed the ozone layer shifted back to them and sales of nonaerosol CFCs soared to new heights. The reasons for this turnabout in fortunes were varied and complex. Politics, economic pressures, and scientific confusion were all to blame.

To bring about additional regulations on CFCs used in refrigerators, air conditioners, foam packaging, and other industrial applications, scientists would be required to prove all over again that the chemicals were depleting the ozone layer. And this time more evidence would be needed, the kind of proof that no one wanted to see: actual measurements of ozone depletion.

To industry, this ultimate proof was considered an impossibility. Industry officials simply didn't believe that ozone depletion could occur in their lifetime. They apparently gave little consideration to the kind of earth their grandchildren would inherit.

It was an easy time to adopt a wait-and-see attitude. By the time EPA officials got around to addressing regulations for all CFCs, in the late 1970s, the climate toward regulations was weakening. And, by late 1980, the new Reagan administration moved into Washington, favoring limited government involvement in environmental matters. The administration emphasized that ozone depletion was a global problem requiring international cooperation. The U.S. aerosol ban had covered 25 percent of world CFC use. But, despite its presence as a global power, U.S. officials became increasingly reluctant to act on further restrictions which could cripple the economy while other nations blithely ignored the problem.

The idea that an international agreement be a first step toward CFC regulations was encouraged by CFC manufacturers. While the aerosol industry was burned by the ban, other CFC industries grew up very quickly in the late 1970s, becoming more sophisticated and politically savvy in their efforts to halt additional CFC regulation. Some industries were particularly shrewd. For several years, the automobile industry had worked hard to separate itself from the aerosol industry so that it might avoid being included in a ban. These CFC users like to point out that the phase out of the so-called "essential" CFCs was a completely new ball game. These were not convenience items that could easily be replaced. Who could imagine life without air conditioning and refrigeration? And what chemicals would take their place? In the late 1970s, there were few. CFCs, for example, were the only known chemicals that could be used to clean telephone switch gears— an obscure responsibility, perhaps, but crucial to the nation's ability to reach out and touch someone. But industry kept discussions of possible alternatives quiet. Home air conditioning units, for example, used CFC-22, a much less dangerous chemical to the ozone. But the automobile industry was reluctant to discuss CFC-22 as a potential replacement for CFC-12 (an ozone-depleting chemical) for auto air conditioners. Changing coolants would cost money, and automobile manufacturers were trying to cut costs, not add them.

The period following the aerosol ban was referred to as the "dark years" by Alan Miller, an attorney at the Natural Resources Defense Council who inherited the CFC-ozone issue from Karim Ahmed and was one of the few who kept the flame burning for additional CFC regulations. But in the early 1980s, even Miller began to have his doubts about whether the issue could be revived. Other environmental groups had dropped the ozone issue, and even his own agency had decided to divert its energy and resources to other problems. Miller, a young, dark-haired attorney who had an air of intellect about him (he later became a law school professor), could see that there were several factors working against the issue. One, of course, was the new administration's disinterest in the matter. The Reagan administration's policy, however, rested on the most current scientific consensus regarding the CFC theory. And the theory was up in the air.

The theory seemed to be trapped in a revolving door. The more scientists learned about chemical reactions in the upper atmosphere, the more complex the issue of ozone depletion became.

During the years from 1976 to 1984, estimates of ozone depletion fluctuated wildly from almost 20 percent depletion to no loss at all. Somehow, the predictions always seemed to float back to the 7 to 13 percent depletion that Rowland and Molina had postulated in 1974. That fact was a small comfort to the Irvine chemists as they watched production of nonessential CFCs increase to the same levels that existed before the aerosol ban. While public concern over ozone depletion faded, they felt that the problem certainly had not.

□ □ □

"The problem with ozone is that no one knows how much of a problem it is," began a story in a news magazine on November 25, 1978.

The confusion over the CFC-ozone hypothesis did not result from any lack of effort among scientists, however. Following the NAS' 1976 report, there was a distinct feeling of urgency as atmospheric chemists attempted to identify all the players in the stratosphere. Now that they had a glimpse of this unique chemistry, they were driven to discover how this previously overlooked part of the world's climate functioned. Throughout the late 1970s and 1980s, their efforts were remarkable.

The computer models used to calculate chemical reactions were dramatically improved as more detailed knowledge about the rates of particular chemical reactions were figured in. In use since the early 1960s, the models were poor substitutes for actual measurements of the atmosphere. Atmospheric models try to duplicate climate conditions by using millions of calculations and the laws of thermodynamics and motion. But there was simply no way, in the late 1970s, to account for all of the variables that might churn out an accurate picture of the world's climate. Despite increasing sophistication, computer models are even today called "dirty crystal balls."

So, even as the models and calculations became more accurate, the addition of more chemicals to the models led to a muddier picture overall. It was like trying to shovel snow during a blizzard; just when the picture was becoming clear, new data would come in.

In an attempt to understand the complete chemical soup of the stratosphere, scientists began focusing on the many other compounds besides CFCs that might have an effect on ozone. For example, atmospheric chemists knew from the SST debates that nitrogen oxide released by the decomposition of nitrous oxide could deplete ozone using the same type of catalytic chain reaction as occurred with chlorine. But there were many more chemicals to consider. In the 1930s, only four primary chemical reactions were recognized in the upper atmosphere. By the 1980s, hundreds had been introduced into the computer models.

Chemists focused most intently on increases in the atmosphere of chlorine, nitrous oxide, carbon dioxide, methane, nitrogen oxide, carbon monoxide, and hydrogen fluoride. Many of the new chemicals were free radicals which are highly unstable and reactive. Appearing and disappearing quickly, such chemicals are almost elusive and are extremely difficult to measure. And yet, just as a quarter teaspoon of the right spice can make a gourmet dish delectable, each chemical in the atmosphere, however minute in quantity, had some effect on the brew that ultimately balanced the ozone layer. Each time a new chemical was introduced, and its rate of reaction figured in, the modeler calculated another estimate of total ozone depletion. And yet almost every scientific report or paper that followed the introduction of a new chemical reaction referred still to "missing chemistry" that might alter the picture again.

Inevitably, scientists reached the conclusion that altitude plays a key role in determining what reactions take place in the atmosphere and to what degree ozone is affected.

As far back as 1974, scientists recognized that reactions in the upper stratosphere, where ultraviolet light is most intensive, are different than those that take place in the lower part of the stratosphere, where less ultraviolet light is available. Even the same chemical could react differently depending on where it was dissociated. As a result, certain chemicals were likely to increase ozone in the lower stratosphere while decreasing ozone in the upper stratosphere. The phenomenon became known as the "compensating factors" in ozone depletion calculations; the end result was a strange sort of balance sheet in which total ozone might remain the same but the earth would still be the loser. The reasons were simple.

While increasing amounts of ozone in the lower stratosphere might effectively block excessive ultraviolet light from reaching the earth, the redistribution in ozone could alter the temperature of the stratosphere, and thus influence world climate.

To Sherry Rowland, this concept was of extreme importance. And yet, because some computer models showed that total ozone might not change, it was an environmental problem that could be easily overlooked, especially by policy makers. As the years went by, Rowland and Molina found themselves in the peculiar position of being essentially right—CFCs did deplete ozone—but being judged to be wrong because the total amount of ozone might not change. To Rowland, the fact that a lot of ozone was being lost high in the stratosphere did not offset the fact that ozone was not being lost below. There would still be climatological effects from this redistribution, he argued. Crutzen, also, was convinced that the redistribution of ozone was important.

Nitric oxide was one chemical that affected ozone, depending on where it was released in the atmosphere. But, like many chemicals scientists examined, the total amount of nitrous oxide being released into the atmosphere was very difficult to determine. Unlike manmade CFCs, whose total quantities produced could be reasonably estimated, nitrous oxide resulted from both natural and unnatural sources. Released by bacteria in the ground over thousands of years, nitrous oxide plays an important role in the natural balance of ozone. But the addition of nitric oxide from unnatural sources, such as farming and aircraft exhaust, transformed the picture. As Johnston pointed out in the early 1970s, nitric oxide released at higher altitudes, where supersonic aircraft fly, could deplete ozone. But nitric oxide released from subsonic aircraft could increase ozone through a series of chemical reactions. This occurs when chlorine monoxide reacts with nitric oxide. An oxygen atom in the chlorine monoxide is shifted to the nitric oxide producing a free chlorine atom and nitrogen dioxide. When nitrogen dioxide absorbs sunlight, an oxygen atom is released. That oxygen atom can be used to create more ozone.

Each new scientific report on ozone depletion from 1977 to 1985 seemed to add a major new chemical reaction. Besides factoring in nitrous oxide, a 1982 report by the National Academy of Sciences on the status of the CFC-ozone theory explored the release of carbon dioxide in the atmosphere. Increasing amounts of carbon dioxide from the burning of fossil fuels, the committee reported, would contribute to a rise in temperature of the lower atmosphere—the so-called greenhouse effect. But the chemical might also cause a decrease in temperatures in the upper stratosphere. This, scientists reasoned, would result in less ozone depletion because it would accelerate the chemical reactions that make ozone and slow the reactions that deplete it. Again, while the ozone layer is spared, the earth is the loser because of the greenhouse effect.

The greenhouse effect was a concept that was formulated early in the century and is considered one of the least controversial theories in atmospheric science. But up until the early 1980s, the theory had received very little attention. In the greenhouse effect, gases migrate to the earth's lower atmosphere and form a kind of dome over the planet. This layer of gases traps the sun's energy at the earth's surface, leading to an eventual warming of the lower atmosphere and changes in air circulation and climate. The theory is not very controversial because scientists are able to calculate fairly easily the radiation absorption potential of various gases.

The greenhouse effect is what allows for life on earth, and yet, could destroy life. Billions of years ago, the earth's atmosphere was lethal for life forms. The atmosphere was mostly carbon dioxide with very little oxygen, hydrogen, or water. But life emerged through these hostile conditions, and the resulting mix of gases and water allowed for just enough warming to sustain life. Too little greenhouse warming and the earth would begin to look like Mars—a planet without a protective greenhouse covering. Too many gases and the world would overheat.

The carbon dioxide that resulted from industrialization (the burning of fossil fuels) has long been recognized as a contributor to the greenhouse effect. But in 1975, a young scientist named Veerhabadrhan Ramanathan, working at NASA's Langley Research Center, discovered that CFCs were highly effective absorbers of the sun's radiation and could contribute significantly to the greenhouse effect.

Ramanathan's discovery was a direct offshoot of Rowland and Molina's work. Early in 1975, Ramanathan had called Cicerone to ask for some help and mentioned that the long survival times of CFCs, as shown by Rowland and Molina, had caught his eye. Like Cicerone, Ramanathan spent the next decade arguing that the greenhouse effects of CFCs added to the reasons for regulating the gases.

Other major contributors to the greenhouse effect are methane and nitrous oxide. But the rate at which the world climate would warm and the actual effects of that warming were a mystery to scientists writing the 1983 NAS report. The atmosphere was known to be notoriously changeable—easily affected by fluctuations on the land and oceans. For example, some scientists predicted that increases of carbon dioxide could increase evaporation of water vapor from the oceans. Water vapor, a greenhouse gas, could contribute to more warming.

Others said that the oceans' ability to absorb heat might mask or delay the eventual effects of global warming for an undetermined amount of time. Cloud cover would also change, but no one could say whether it might increase or decrease global warming.

These feedback mechanisms were not (and still aren't) well understood. James Lovelock, who also recognized that CFCs were a potential contributor to the greenhouse effect, had suggested that the earth might even compensate for these changes. Lovelock's controversial Gaia theory, which he published in the 1970s, suggested that the universe is driven to a state of balance and organization. The earth's living organisms are able to affect the atmosphere and environment in order to protect life. While boasting that Gaia was enormously resilient, Lovelock later conceded that manmade pollutants on the order of what was being emitted in the 1980s could easily overpower any Gaian system.

In the early 1980s, however, the greenhouse effect seemed like some kind of possible, distant threat. While the scientists working on the 1982 NAS report recognized the threat, they were apparently stymied by what they saw as a lack of proof on the accuracy of all of the chemical reactions that contributed to atmospheric change.

In the face of this doubt, it was incumbent on atmospheric chemists to spend as much time trying to actually measure ozone and chemicals in the atmosphere as they spent before their computer models. Measurements of total ozone had been of interest to scientists for several years due to an unusual increase of ozone in 1960s. Now, however, ozone measurements might allow scientists to detect an early warning that overall depletion was occurring. Finding particular chemicals at various altitudes would help assure that the information going into the computer models was correct—that certain reactions were indeed taking place. But this proved to be easier said than done.

☐ ☐ ☐

For Jim Anderson, a University of Michigan physicist, detecting the presence of particular chemicals was like trying to spot "a drop of vermouth diffused in an Olympic-sized swimming pool full of gin." Anderson was a patient, hard-working man with a very soft voice and a good sense of humor. He was determined to try to measure the concentrations of chemicals in the atmosphere. His contributions to the evolving CFC-ozone hypothesis were both significant and infamous. In late 1976, Anderson became the first person to find evidence of chlorine oxide in the stratosphere.

Detecting the chemical would provide the third, and supposedly final, piece of evidence needed to show that the catalytic chain reaction Rowland and Molina had proposed was actually taking place. Hydrogen chloride, which the models predicted would be present if the theory was correct, had already been detected and was shown to increase with altitude. This meant that chlorine atoms probably were released in the atmosphere from a reaction. The "smoking gun," then, was the presence of chlorine or chlorine oxide. Chlorine oxide could only be produced by the action of chlorine and ozone, but the chemical was difficult to measure since it wouldn't be present in large amounts. Anderson did detect the chlorine oxide. Unfortunately for everyone involved in trying to confirm the hypothesis, however, he found too much.

Detecting the chemical wasn't easy. In late 1976, Anderson conducted a series of tests using special scientific balloons to look for chlorine oxide. Despite the importance of Anderson's tests, the setting was far from what one would envision as a location for sophisticated stratospheric chemistry. The National Scientific Balloon Facility was located off narrow, dusty roads in the tiny East Texas town of Palestine. A low-budget program operated by the National Science Foundation in the mid-1970s, the facility consisted of a cluster of white buildings near some farm land. But it attracted teams of scientists from prestigious universities and research labs around the country. High-altitude balloons were, Anderson felt, the overlooked stepchildren of astronomy and high-energy physics. Described by many as economical gasbags, the balloons were made of acres of microthin polyethylene film and were often 800 feet long. Filled with helium, the balloons towed sensitive, expensive, scientific equipment—minilaboratories, actually—to a maximum of 30 miles above earth. Despite being able to soar so high, the balloons were incredibly gentle, thus allowing measurements to be taken with surprising sensitivity.

Anderson and his research team were looking for a rather simple and fundamental interaction between light and the electrons contained in these free radicals: atomic chlorine and chlorine oxide. The experiments were designed to be compared to what the models were calculating at the time. Having the models predict the presence of chlorine oxide was one thing, but an observation, Anderson knew, would be the proof everyone was looking for.

After extensive and tedious laboratory work to prepare the instruments for the balloon flights, Anderson's team was ready to launch their traveling chemical laboratory. Once the balloon was situated, the scientists would cut the instrument off the balloon and it would plummet at high velocity back to earth on a parachute, taking measurements along the way. The entire window for gathering measurements was only five minutes. And within hours, Anderson's team was able to process and study the results.

It was clear from the very first measurements that both atomic chlorine and chlorine oxide were present in the stratosphere. In fact, in Anderson's first seven flights, the measurements of chlorine oxide were twice what models had predicted, a value of 2 parts per billion instead of 1 ppb—a finding that didn't upset anyone greatly. But the eighth flight, which Anderson conducted on Bastille Day (July 14) 1977, threw his entire body of work into question. That day, Anderson's experiment showed chlorine oxide at the unbelievable magnitude of 8 ppb.

The news of Anderson's bizarre Bastille Day measurement sent the field into a frenzy. The odd result raised questions about the validity of Anderson's earlier chlorine oxide measurements. And, without credible chlorine oxide measurements, the whole ozone depletion theory, which had been gaining support, was in doubt again.

Part of the hysteria over Anderson's odd measurement, Rowland felt, had to do with the publicity surrounding the CFC-ozone argument. In a laboratory experiment, such an odd result would be looked at as a mere anomaly. And, if it didn't appear again—and Anderson's odd measurement was never repeated—it would be ignored. Most experiments were complicated and an anomaly is difficult to diagnose. Often, unusual results were the consequences of equipment failure. But ozone depletion was a highly public issue in 1977 and, said Rowland, "You feel as if you need an explanation for it."

Anderson, in particular, fretted over what he called his "black sheep" value. Sensitive to the controversy and inspection his series of experiments had generated, Anderson did his best to come up with an explanation for the odd result. He and his coworkers had spent a great deal of time trying to understand their instrument and tended to believe that the high value was real because the instrument had been so carefully and sensitively calibrated. It had been cross-checked numerous times. What's more, Anderson felt it was simply unethical to just throw the odd result out—pretend it didn't exist. Some of his colleagues didn't understand this reasoning, however.

"It's so far outside the normal band of your observations, why don't you just ignore it?" well-meaning friends asked him.

Instead, Anderson came up with the only explanation that made sense. He concluded that the measurement may have resulted from high chlorine oxide produced by a disturbance in the local atmosphere such as a meteor or passing rocket releasing gases. As bizarre as the explanation seemed, it was theoretically possible. To those who suffered through 15 years of the ozone issue it seemed sort of typical, in fact, that a wayward rocket might come along at just the right moment to muck things up further. Cicerone and Rowland found the obsession over the odd measurement especially ironic. For three years, people believed that detecting chlorine oxide in the stratosphere would confirm ozone loss. Instead, finding the substance had led to more doubt.

No one was able to actually prove what had caused the Bastille Day measurement. And the issue was eventually forgotten when researchers at the National Center for Atmospheric Research found measurements closer to what models predicted. Scientists at the Jet Propulsion Laboratory also found values of 2 ppb, higher than the hypothesis predicted but nowhere near the odd result Anderson had found. Anderson himself delved back into his experiment with more balloon flights. And the measurements that followed strengthened the body of the work and confirmed what others were finding.

□   □   □

The detection of chlorine oxide, however, didn't convince everyone of the validity of the Rowland-Molina hypothesis, as the Irvine researchers had hoped it would. Rowland later called the initial CFC-ozone depletion theory the "first early warning" of potential atmospheric disaster and the discovery of chlorine oxide the "final early warning." If more proof were required after that, he said, it would probably not be in the form of a warning but would be actual measurement of ozone depletion.

And, by the late 1970s, it seemed entirely possible that ozone depletion would have to be observed before anyone would do anything about CFCs.

In a 1979 statement, Du Pont officials stated: "No ozone depletion has ever been detected despite the most sophisticated analysis. . . All ozone-depletion figures to date are computer projects based on a series of uncertain assumptions."

Part of the ·problem was the lack of any good way to measure total ozone. Dobson stations, ground-based ozone monitoring instruments, had been in place at various locations around the world since the 1930s with the numbers of stations increasing greatly in the 1950s and 1960s. No one was sure just how accurate the instruments were, however. There was also the question of whether ozone loss could even be detected as yet since CFCs could survive a hundred years in the stratosphere. Finally, any ozone loss from unnatural sources would have to be distinguished from the natural ozone fluctuations that occurred with the 11-year solar cycles. When sunspots are fewest during this cycle, ultraviolet light is weakened and less ozone is produced. How to detect this thin footprint of unnatural ozone loss from the natural backdrop was a source of continuing frustration for atmospheric scientists.

Donald Heath, a scientist at NASA's Goddard Space Flight Center in Greenbelt, Maryland, was convinced that satellites could best measure total ozone. Heath, a short, intense man who wore glasses, had been measuring ozone from research satellite data since 1970. In 1978, he had announced that data taken from the Nimbus 4 satellite found only half the ozone depletion detected by ground-based Dobson stations. The discrepancy, Heath said, could be cyclic, short term, long term, natural, or manmade. Take your pick. Better measurements, he said, would soon be available from the Nimbus 7 research satellite which was launched in 1978.

But further analysis of satellite data only added to the confusion. By 1981, Heath and his colleagues at Goddard felt confident enough of the accuracy of the satellite readings to announce preliminary evidence that the ozone layer was gradually being depleted. Heath took data from two satellites—Nimbus 4 and Nimbus 7—and analyzed it to obtain some type of picture of ozone levels. According to the analysis, ozone appeared to have decreased by a total of 1 percent at an altitude of 25 miles during the years between 1970 and 1979. Heath said the results could not be directly attributed to the presence of CFCs but pointed out that if CFCs were damaging ozone the effects would be seen at 25 miles altitude.

"There is no way that can happen fortuitously," Heath told *Science* magazine. "Something real has happened in the atmosphere."

Heath submitted the paper to *Science* but failed to get it published. The study, many felt, was simply too controversial. There were too many questions about the accuracy of comparing one satellite's data with another's, and neither his superiors at NASA nor the Chemical

Manufacturers Association would believe it. But, Heath protested to
his colleagues, the data was not presented as proof of ozone depletion.
He merely saw the study as an indication that things were happening
with the ozone layer.

"It was sort of the first time anyone had made such a study to
pick out such a small effect," Heath said later. "There were many
questions, but I still believed it was real. There was so much opposition
to it that I sort of let it die. I thought I'd wait awhile."

While Heath's findings were questionable, they were still the first
suggested evidence that ozone was being lost around the world. But
times had changed. Scientific evidence was only one small part of a
highly political argument on protection of the ozone layer. At the EPA,
Herbert Wiser of the agency's Office of Research and Development,
called Heath's findings "mildly suggestive" of global ozone loss.

"What we don't know is the magnitude of the effect," Wiser told
Science News. "If the magnitude is large, we need regulatory action.
If the magnitude is small, we may do nothing."

With each new bit of research, government agencies dutifully summarized what was known about ozone depletion and made predictions. There were no lack of studies on the problem. But the constant
analysis only seemed to lead to greater confusion. Robert Watson, a
NASA scientist who directs the agency's studies of the upper atmosphere and who was to become a key participant in major ozone
discoveries of the 1980s, later told *Discover* magazine that the ozone
studies "all fundamentally said the same thing, but what most people,
especially policy makers, were doing was spending their time analyzing the differences between them rather than the consistencies."

This trend toward magnifying the uncertainties regarding ozone
depletion began in 1979 when the National Academy of Sciences issued
its second report on the issue.

While the September 1976 report had put ozone loss at 2 to 20
percent with a likely loss of 7 percent, the new study was said to be
based on refined models and the new values for rate constants. Panelists concluded that if CFC output continued at current world rates,
ozone would be depleted by 16.5 percent by late in the twenty-first
century. The panel members suggested that a "wait-and-see" approach
to the controversy was not practical and that the United States should
take the lead in pursuing international cooperation to control CFCs.

The impact of the NAS report was quickly blunted, however, with a study by the Stratospheric Research Advisory Committee conducted for the United Kingdom's Department of the Environment. The UK study concluded that the ozone-depletion hypothesis was in doubt because of inaccuracies in computer models and the lack of understanding regarding chemical reactions in the stratosphere.

The discrepancies boosted the CFC industry's hopes to delay further regulations. In the offices of the Senate environmental committee, staffers met with Du Pont representatives to discuss the latest NAS report. But Du Pont officials attended the meeting with binders containing excerpts from the British report. The differences between the U.S. and British reports, they charged, were evidence of a lack of scientific consensus on the matter.

Du Pont's new argument incensed Curtis Moore, an environmental committee staffer who had worked for the Senate committee during the SST debates. Moore suggested to Senator Robert Stafford of Vermont that they write to a British scientist who had worked on both reports and ask him if there was a scientific consensus on the amount of expected ozone depletion. The British scientist answered the query and explained that the U.S. report was probably the more accurate because scientists used two-dimensional models, which showed ozone at various altitudes as well as various longitudes and latitudes, while the British scientists had used the simpler, altitude-only one-dimensional models.

Moore, a self-described cynic who cared deeply about the environment, also kept in touch with the EPA on its proposed Phase Two regulations. But he, like others, sensed that the ozone issue was dying. EPA officers had promised that the new NAS report would influence their actions on additional CFC regulations. But, despite the fact that the nation's leading scientists had doubled their estimate of ozone depletion, the report, said one science writer, had "fallen on deaf ears." Signs that regulations on nonessential uses of CFCs would be subject to greater forces than science had been evident for months.

Almost as quickly as bans on CFCs were put into place, federal agencies had begun to back down on additional regulations. After enacting Phase One of the ban on CFCs for nonessential products, the EPA, FDA, and CPSC announced their intentions to draft regulations for a Phase Two ban on nonpropellant CFCs by June 1978. On January 13, 1978, the three agencies called for a public meeting to discuss "the need" to regulate nonpropellant uses. At the February meeting, a decision was made to delay regulations due to the complexity of the issue.

The scientific uncertainties had contributed to this delay. But what was perhaps more influential in the decision to postpone additional regulations were the political and economic realities.

The economic problem was the high price of substitute chemicals for CFCs should they be banned. CFC manufacturers had initiated research into substitutes for the chemicals used in refrigeration, air conditioning, foam blowing and cleaning solvents in the mid-1970s. By 1979, several substitute chemicals had been identified but none could be produced as cheaply as CFCs 11 and 12. They had, after all, been in production for 40 years. By the early 1980s, with Phase Two regulations seeming more and more unlikely and the public no longer interested in the ozone depletion issue, most manufacturers suspended their efforts to even find substitutes without having done the toxicity tests that are required of new products and that take up the bulk of time in advancing a new product to the market. At Du Pont, officials dropped their search for CFC substitutes citing the EPA's lack of interest in the issue and because CFC substitutes would be costly and difficult to make. Without the pressure of pending government regulations, one high-ranking Du Pont official later recalled, there seemed no point in pursuing the research on substitutes. Who needed substitutes when it looked as if CFCs were going to be around for awhile?

"No matter how good a material is, if no one will buy it, what good is it?" said Donald Strobach, director of Du Pont's Environmental Division. "The (research) program came to an end because there wasn't enough interest."

In hindsight, Mario Molina, who left the University of California, Irvine, in 1982 to work for NASA's Jet Propulsion Laboratory, felt that the lack of research on substitutes was the biggest error in the ozone war. Molina had kept in touch with Du Pont officials in order to discuss progress on substitutes. But the calls and discussions were futile.

"In retrospect, the mistake was (for government) not to provide the incentives to develop alternatives," Molina said. "It was very clear to Sherry and I that the risk (of not finding substitutes quickly) was very large. This was a sort of failure on our part to make that even clearer. Why take this risk? For society, it's not a major investment to try and look for these alternatives. We were not advocating suddenly to stop air conditioning or to stop home refrigeration. That need not be done at all. It was a matter of starting sufficiently ahead of time to have a smooth transition."

It was to become a failure on the part of CFC companies' research directors that the research was stopped. Had one company pursued alternatives, it could have positioned itself far ahead of its competitors when CFC regulations were ultimately announced. None of the research directors in the CFC industries apparently saw the likelihood of a market for alternatives that, in 1988, would be estimated at $1 billion. A company with a CFC substitute handy could also have reaped a financial bonanza with licensing agreements.

But, in the early 1980s, industry not only bemoaned the lack of safe, nontoxic substitutes but dwelled on what it called a dearth of hard scientific data proving the hypothesis. From 1974 through 1987, industry kept asking for "an additional three years" to do more research. (Molina called the tactic the "sliding three years.") And, perhaps more effectively, industry argued vehemently before policy makers that it was unfair to phase out CFCs in the United States while other countries continued to enjoy the practical and economic benefits of using the chemicals. While U.S. officials had hoped other nations might follow its lead in banning CFCs in propellants only Canada, Sweden and Norway followed with similar bans.

Both the EPA and the State Department had tried to initiate international talks on the CFC-ozone issue. Although talk on worldwide environmental problems was new, some groundwork had been lain prior to the CFC-ozone controversy to help deal with such matters.

In 1972, the United Nations had led a series of international scientific conferences that culminated in the UN Conference on the Human Environment in Stockholm. Besides creating an office called the United Nations Environmental Programme (UNEP), the 1972 conference opened the door for international action on environmental matters. Prior to this, international agreements on environmental matters were limited to the protection of natural resources or wildlife. The Stockholm conference, however, led to the adoption of an international toxic substances principle stating that a nation must take responsibility to ensure that its activities do not cause environmental harm beyond its own jurisdiction.

The United Nations Environmental Programme became an early leader in opening discussions on the global CFC problem. Two international meetings on ozone depletion were held in 1977. The first, in Washington in March, was sponsored by UNEP to discuss the evidence for ozone depletion and the need for more research. The meeting was attended by the European Economic Community and 33 nations. The presence of EEC officials was crucial because the block produced about one-third of the world's CFCs and dictated conditions that might affect international trade. Two more meetings followed in 1978 with participants agreeing that efforts toward CFC regulations should begin. But a commitment to eliminate the chemicals was superficial. While the EEC eventually agreed to put a ceiling on CFC production, it was far above what the countries were actually producing.

In a major meeting in Oslo in April 1980, however, representatives from the United States, Canada, Sweden, Norway, Denmark, the Federal Republic of Germany, and the Netherlands agreed that waiting for additional proof of ozone depletion was dangerous and called for all nations to immediately begin reducing CFCs from all sources. EPA officials went a step further, announcing the United States' intent to freeze CFC production at 1979 levels.

"We in the United States are taking this step because of continuing studies showing that worldwide chlorofluorocarbon emissions jeopardize public health and the environment," Barbara Blum, EPA deputy administrator told the negotiators.

Blum cited the 1979 NAS report as justification for the EPA's action and suggested that the ceiling would be neither the first nor last step to control CFCs.

Despite the sometimes noble talk at international meetings, it was obvious that any international agreements to limit the chemicals would take years. Too often, international agreements were made before individual nations could guarantee their own country's support for domestic regulations. Negotiators knew that CFC regulations would be particularly hard to implement because of the degree of scientific uncertainty. The leaders of many nations were simply not convinced of the dangers of the chemicals and, thus, felt they had little reason to act.

International negotiations too often missed the point, experts observed. According to a 1982 study by researchers at the Center for Technology, Environment and Development at Clark University in Worcester, Massachusetts, negotiators at the ozone meetings were stymied by the fact that ozone depletion represented a slow change. It was a change that might not even be apparent for a century, was easily disguised by natural depletion, and was, therefore, steeped in scientific uncertainty. Various nations' responses to the problem were based on widely differing attitudes on environmental matters, approaches to decision making, and what was at stake, economically, regarding the issue. In the face of doubt, international negotiators clearly relied on values other than science to balance the risks and benefits of regulating chlorofluorocarbons.

Little was to come from the largely symbolic meeting in Oslo. But in the United States, the EPA's proposal to freeze CFCs at 1979 levels set industry in a state of near panic.

In the summer of 1980, industries using CFCs, along with some CFC producers, responded to the threat by forming the Alliance for Responsible CFC Policy, a trade group charged with lobbying against additional regulations. The group's purpose, an industry spokesman told the *New Yorker*, "was to convince the government—Congress, the White House and anyone else—that EPA's proposal to restrict CFCs is ill-advised."

In October of 1980, during the waning days of the Carter administration, the EPA officially released its proposal for the phase-out of nonessential CFCs. Called the Advance Notice of Proposed Rulemaking (ANPR), the EPA suggested two approaches. The first was a mandatory-controls approach to place a cap on CFCs through restrictions on production. The cap would consider what was technologically feasible at the time. Thus, the EPA could ban some uses of the chemical, demand recycling of CFCs in some uses, or ask that alternative chemicals be used.

The second approach was an economics-incentive tactic that would call for a ceiling on production through a system of permits according to price. The agency proposed a limit of 30 percent of the present levels with a cap on production and a timetable to lower the cap each year. A notice was published on the proposals asking for public comment, and EPA officials sat back to wait.

For the first time in years, there was a significant consumer response to the CFC-ozone issue. But these were not the same consumers who cried "ban the can" years earlier. Of a reported 2,300 letters received by the EPA, only 4 supported additional CFC restrictions. The other 2,296—undoubtedly CFC users—fiercely opposed the regulations.

7
A Crisis that Wasn't
January 1981—March 1985

By 1981, the CFC industry was beginning to breathe a bit easier. There was an almost audible sigh of relief in May 1981 when Anne M. Burford, nominated by President Reagan to head the EPA, testified during her confirmation hearings that she thought the CFC issue highly controversial and that there was a "need for additional scientific data before the international community would be willing to accept it as a basis for additional government action." In her 1986 book *Are You Tough Enough?*, Burford refers to the ozone issue as a scare issue that she was savvy enough to dismiss as unimportant. "Remember a few years back when the big news was fluorocarbons that supposedly threatened the ozone layer?" she wrote.

Burford's brief reign over the EPA was a bleak era for activists seeking CFC controls and ozone protection. At the NRDC, Alan Miller was continually frustrated in his attempts to get Burford's attention on the CFC-ozone matter. One problem was that Burford reportedly said she would refuse to meet with any environmentalists during her first year in office. But when Miller finally did get permission to set up a meeting between environmentalists and Burford to discuss the matter, the ozone depletion issue was so dead in Washington that he had to beg old friends who owed him favors to show up at the meeting and lend support.

Under Burford, the EPA tried to delay domestic regulations on nonaerosol CFC uses while pushing the State Department to develop an international plan for reducing the chemicals.

The EPA and State Department efforts were channeled through

the United National Environmental Programme which had, in May 1981, established a Working Group to try to come up with a global agreement—called a convention—for protection of the ozone layer. The group met seven times from 1982 to January 1985. But until Burford left the EPA in 1983, the United States agreed with its European allies that an international agreement on ozone should begin with a framework convention—to specify the rules for future discussions— followed by a protocol specifying regulations.

Burford's view that the ozone issue was simply not an urgent matter was backed by the administration. In letters sent to the administration's deputy secretary of energy, Miller was repeatedly and cheerfully put off. The President's cabinet seemed to have the utmost confidence that science would prevail and save the world from any peril caused by chemicals.

The Reagan administration had appeared to have blind faith in the nation's scientists and technology when it came to pollution matters. When an environmental report entitled the *Global 2000 Report* was released warning that pollution problems would endanger the world's natural resources and human health by the year 2000 if controls were not undertaken, the administration responded with a campaign that humans were the "ultimate resource." It was as if science could find a way to undo any damage wrought by the earth's inhabitants. The idea of preventing environmental catastrophe apparently didn't occur to anyone in the White House during that brief but unhappy era.

The Reagan administration was also sympathetic to the idea that additional CFC regulations could cause grave economic hardships. By the early 1980s it was apparent that substitutes for the "essential" CFCs would not be as easy or as cheap to find as aerosol substitutes were. In the summer of 1981, with Burford ensconced in EPA Headquarters, an EPA official went before the House Subcommittee on Antitrust and Restraint of Trade Activities in a meeting to consider the effect of CFC restrictions on small businesses. The agency and the White House had been besieged by calls from CFC manufacturers and related businesspersons who were horrified with the EPA's announcement of a freeze; some of them mistakenly thought that the EPA had already decided to implement additional bans. Edward A. Klein, director of the EPA's Chemical Control Division in the Office of Pesticides and Toxic Substances, said that the EPA's October 1980 notice did not mean the EPA had decided to implement additional CFC regulations.

"Let me assure you that EPA is extremely sensitive to the needs of small businesses and that these needs will be carefully considered, not only with respect to CFCs, but for all matters addressed by EPA," Klein said in his testimony.

Klein's reassurances were not enough for Thomas Luken, chairman of the Subcommittee on Antitrust and Restraint of Trade Activities Affecting Small Businesses. Repeatedly urging the EPA to consider the needs of small businesses and accusing agency officials of already having their minds made up, Luken questioned Klein relentlessly.

Luken: "In your statement, you indicated that the ANPR—the Advance Notice of Proposed Rulemaking—is not a rule. I believe you say it is not even a statement that EPA will regulate CFCs. . . Since you have indicated what it isn't, is it nothing?"

Klein: "It is an information-gathering tool. As my testimony indicated, we got—."

Luken: "Well, you wouldn't issue it if you weren't thinking about it just slightly, would you?"

The Reagan administration viewed the ANPRs, which were issued in the final days of the Carter administration, as evidence of the democratic administration's tendency to overreact and overregulate. And the July 15, 1981, hearings provided the CFC industry with a forum to make its strongest case against new regulations. In a statement to Luken's committee from the industry lobbying group, Alliance for Responsible CFC Policy, manufacturers raised seven key points to argue against the EPA's proposed regulations. The Alliance claimed:

☐ Its Ozone Trend Analysis, in which ground-based ozone-monitoring stations were periodically reviewed, would provide an early warning system for any developing problem in the stratosphere.

☐ Ozone depletion had not been measured.

☐ The 1979 NAS reports cited by the EPA as justification for future regulations had been subject to serious criticism and that actual ozone depletion would only be about one-half of that stated in the report.

□ New scientific studies would reduce ozone depletion predictions even further.

□ Atmospheric science was in a period of very rapid change.

□ The current status of the CFC-ozone depletion theory did not justify further regulations of CFCs.

□ There was a negligible risk in waiting for the results from further research.

□   □   □

The Alliance also called for an international consensus on ozone depletion before any U.S. action was taken. It charged that the EPA "virtually ignores the beneficial aspects of CFCs" in its Advance Notice for Proposed Rulemaking and accused some agency members of having made up their minds to push for new regulations.

"EPA personnel who have prejudged the issues must be disassociated from the proceeding," the Alliance stated. "In any event, EPA must take a fresh look at all evidence submitted in this matter. The agency cannot rely upon the 1978 aerosol propellant ban, which does not constitute a factual or legal precedent or basis for further regulation of CFC uses."

Although it seemed to be one tiny voice shouting against a chorus of dissenters, the Natural Resources Defense Council responded to the hearings and the Alliance statement with an angry statement of its own—submitted for the record because the NRDC had not been invited to testify at the hearings—claiming that "misinformation and misleading statements were made by several witnesses."

One problem was that the hearings focused on the costs of CFC regulation and ignored the benefits, the NRDC claimed in its statement. The costs of regulations—the businesses that would be forced to shut down; the increased price of CFC substitutes—were much easier to quantify than the benefits. The consequences of failing to control CFC use would not become apparent until substantial injury and destruction had already occurred.

The environmental group also came to the defense of the EPA, saying the agency had gone out of its way to listen to industry.

The Advance Notice of Proposed Rulemaking "was a commendable effort to obtain industry and public comment in advance of an agency decision to regulate, much less a commitment to any specific form of regulation. Instead of receiving praise for seeking public comment, the Notice has been characterized incorrectly as a regulatory proposal and made the target of an industry onslaught unprecedented since the Allies invaded Normandy," the NRDC statement charged.

Although they had already earned sympathy from some lawmakers, CFC manufacturers, united by the 400-member Alliance for Responsible CFC Policy, took to the Hill to lobby for legislation that would halt further CFC restrictions. Industry pushed for legislation that, on its face, appeared to provide for CFC regulations but actually precluded any U.S. regulation until there were actual measurements of ozone depletion directly attributable to CFCs or there was an international agreement for regulations.

Luken and Senator Lloyd Bentsen of Texas introduced bills that sought to amend the 1970 Clean Air Act (which was due for renewal by Congress that September) to allow for more research on the ozone layer and the stratosphere and restrain EPA from regulating CFCs without international consensus. The bills stipulated that the EPA would be required "to use statistical analysis of actual ozone measurements as an early warning system to guard against excessive ozone depletion."

Curtis Moore, the assistant who worked on the Senate Environmental Committee, was incensed by the legislation and blamed the CFC industry.

"Here were people who were willing to say they ought not be required to modify their behavior one scintilla until they had massively disturbed one of the fundamental parameters under which life had evolved on earth," Moore said later. "These groups just got more and more offensive to me."

National Academy of Sciences officers were also worried about the proposed legislation. They realized that if such legislation passed they might be asked to be the impartial judges on the need to regulate and argued that their role should be one of research and evaluation of the scientific data.

Meanwhile, Senator Robert Stafford of Vermont, a calm, unflappable man who was seemingly unafraid to take actions that might shut down an entire industry, countered Bentsen's bill with a different amendment. Eventually, a compromise was reached to continue studying CFCs. But Stafford succeeded in halting legislation that would have prevented regulations until evidence of ozone depletion.

Despite this minor setback, the Alliance worked the issue skillfully. There were rumors that the organization had made campaign contributions to key politicians, and it argued a good game for suspending regulation. CFC manufacturers claimed that, on the economic level, unilateral restrictions on CFCs would be disastrous, especially for small businesses. In the spring of 1981, CFCs were a $500 million industry. The Alliance claimed that 260,000 businesses, most of them small, used CFCs, and that no substitutes were available for most applications. Second, its representatives claimed, the EPA was using data from the 1979 NAS report to support its contention that additional CFC restrictions were necessary. But, said Paul W. Haber, Du Pont's Environmental Manager in the Freon Division, "A number of scientific developments in 1980 sharply throw into question the utility of the (NAS) reports and EPA's ongoing reliance on them." Specifically, Haber and other industry officials claimed that revisions in chemical reaction rates substantially lowered estimates of ozone depletion. (At the time, Heath's findings on a drop in total ozone had been leaked but had not been published and, manufacturers argued, the accuracy of the findings could not be verified.)

□   □   □

Without agreement on the severity of ozone depletion, manufacturers could persuade lawmakers that it was premature to implement regulations. Lawmakers like Burford, then, were left to make the easy choice of protecting the economy at the risk of environmental catastrophe later.

Science had little to offer policy makers trying to resolve the issue. It sometimes seemed that reaching a consensus on the extent of ozone depletion would be impossible. Two additional reports from the National Academy of Sciences—in 1982 and 1984—showed that the stratosphere was still an area of vast unknowns.

The 1982 NAS report, released on March 31, 1982, put ozone depletion at 5 to 9 percent, only half that predicted by the 1979 report. The new figures, the academy claimed, were due to improvements in measurements of certain chemical reactions. This time, for example, the academy had looked at the combined effects of CFCs, nitrous oxide, nitric oxide, and carbon dioxide.

Although the estimates were close to Rowland and Molina's original 7 to 13 percent prediction, the tone of the NAS report seemed to indicate that no crisis was imminent. It stated that there was no evidence of a decrease in ozone directly related to human activity.

Other chemists, however, felt the report did not present a complete picture of what was known. The report did not take into account
NASA scientist Donald Heath's satellite findings on total ozone depletion because they were judged to be preliminary. The study also failed
to consider how ozone might be changing in different parts of the
world. In the early 1980s, atmospheric scientists relied on one-dimensional computer models that calculated ozone values only at various altitudes and predicted only average global depletion. Newer,
potentially more accurate, two-dimensional models reviewed changes
in ozone at various altitudes as well as with longitude and latitude and
showed more ozone depletion at latitudes far north and far south of
the equator.

While seeming to downplay the depletion issue, NAS panelists
did, however, warn that the biological effects of increased ultraviolet
light from ozone depletion might be more serious than predicted. The
scientists concluded that increased UV light could be expected to increase skin cancer, including the most serious form, malignant melanoma. The increase in radiation could also inhibit plant growth, although additional research in this field was needed.

The report had no effect on regulation. Indeed, industry seemed
to have been granted another reprieve. That spring, the Pennwalt Corporation, a major U.S. producer of CFCs, revealed plans for a $10
million expansion of its CFC plant in Calvert City, Kentucky, a significant action because the U.S. CFC industry had spent little on growth
since 1976. As of 1982, scientists still didn't have the kind of data policy
makers required to prove that the amount of ozone depletion caused
by CFCs was serious enough to warrant regulation.

The report failed to settle a key question on whether emissions
of CFCs were growing again. In the early 1980s, Reinhold Rasmussen
of Washington State University, the scientist who had tried to disprove
the Rowland-Molina theory in the mid-1970s, reported that measurements of CFC-12 taken each January in Oregon and the South Pole
had increased every year since 1974. This contradicted industry as-
sertions that suggested CFC output had decreased 20 percent worldwide since 1974.

Rowland's contention that CFCs were growing once again was
virtually ignored when NAS panelists met again in 1983 to update their
knowledge of the CFC-ozone issue. Instead, the addition of new chemical reactions to computer models was deemed to be of greatest importance to the new estimates. Unlike previous NAS reports, this time
the conclusions seemed written to assure Americans that no ozone

crisis was imminent. Or, as one headline in a science magazine pro-
claimed, the new NAS report could be summarized, "Ozone: The Crisis
that Wasn't."

The report, released in February 1984, put eventual ozone depletion at 2 to 4 percent based on new calculations that showed more
ozone would be created in the lower stratosphere while some would
be lost in the upper stratosphere. For example, the report found that
emissions of nitrous oxide into the atmosphere would increase ozone
by 1 percent; methane, by 3 percent; and carbon dioxide by 3 to 6
percent. Overall, the report stated, total ozone might actually rise 1
percent over the next decade. The report cautioned that the redistribution of total ozone might lead to climatic changes while the increase
of CFCs could contribute to the greenhouse effect.

To Rowland and Molina, the report was misleading.

"The academy is saying that it looks like there were compensating
factors," Molina told doubters. "But you will still have a large perturbation of the ozone layer. It is still a major disruption of a natural
system."

Looking at the total ozone loss was a way of "minimizing the
importance" of the issue, Rowland said.

"No one has yet succeeded in developing a scenario in which
the increase of CFCs doesn't decrease upper stratospheric ozone," he
said. "But lots of scenarios that have been developed show this offsetting either in the troposphere or lower stratosphere, so that the
calculations of how much total ozone would change have fluctuated
widely."

The academy's new estimates were met with satisfaction by those
who had long doubted that CFCs could have a profound effect on the
earth's natural systems. James Lovelock, writing in the December 1984
issue of *Environment*, said, "Had we known in 1975 as much as we
know now about atmospheric chemistry, it is doubtful if politicians
could have been persuaded to legislate against the emission of CFCs."

As if he regretted his role in the whole unsavory episode—by
being the first to detect CFCs in the atmosphere—he added, "Wars
do not usually start from a single, isolated incident." But this war may
have been worth it, "even if it was a messy and gaudy way to gain
public support and money for scientific research."

Burford had felt no urge to address the ozone issue because of the
1982 NAS report and other scientific studies that continued to show

---

Summary of National Academy of Sciences Reports on Ozone Depletion

Year: 1976
Estimate: 2–20% depletion (7% most likely)
Comments: Rowland-Molina theory substantiated. More studies required to estimate total ozone depletion.

Year: 1979
Estimate: 16.5% depletion
Comments: A wait-and-see approach regarding regulations of CFCs is not feasible.

Year: 1982
Estimate: 5–9% depletion
Comments: Other pollutants (compensating factors) may reduce extent of total ozone depletion but vertical distribution of ozone is likely to change. Biological consequences of ultraviolet light more severe than first estimated.

Year: 1984
Estimate: 2–4% depletion
Comments: Ozone less vulnerable to manmade pollution than first thought. While total ozone may not change, vertical distribution of ozone is likely to change.

The National Academy of Sciences released four major reports on the CFC-ozone controversy from 1976 to 1984. The widely varying estimates of eventual ozone depletion indicate the confusion over the theory.

smaller and smaller estimates of ozone depletion. She wasn't actually flying in the face of science, but science certainly seemed to match her predisposition on the matter. And there was little political pressure for her to act. The international negotiations were moving at a snail's pace and the environmentalists seemed to have forgotten about the issue. At EPA headquarters, there were rumors that the ANPR was going to be withdrawn.

Burford, however, was unable to make a final decision on the ANPR. She resigned from the EPA on March 9, 1983, under pressure over the EPA's performance and besieged by charges of mismanagement and conflicts of interest. Burford's resignation was viewed as a window of opportunity for a very small group of people at EPA who supported additional CFC regulations. And, under the new, more open-minded EPA director, William Ruckelshaus, the pro-regulation staffers began their efforts anew.

Besides the intra-agency interest in reopening the ozone issue, Ruckelshaus was to face a sudden challenge from environmentalists. At the Natural Resources Defense Council, Miller was fighting mad. By 1984, domestic and worldwide CFC production had returned to the same levels as before the aerosol ban with worldwide growth estimated at 5 to 10 percent (the thing Rowland had been warning of). The growth in CFCs made the aerosol ban, only five years old, look silly.

Alarmed that the issue was dying in Burford's EPA, Miller had decided on a final tactic to refocus attention on the long-forgotten Phase Two regulations. Circulating a memo at the NRDC, Miller detailed his plan to sue the EPA for not following up on its Phase Two regulations. While his colleagues supported Miller's plan, the council's board of directors voted to hold off on a lawsuit for a short time. Ruckelshaus had just been appointed to EPA, and the NRDC folks were as glad to see him replace Burford as anyone in Washington. The council decided to give the new EPA director a grace period. Then they would sue him.

Miller filed a notice on the NRDC's intent to sue the EPA in mid-1984, but this time the agency begged for the NRDC's cooperation and patience. U.S. negotiators were moving ahead in sensitive talks to establish an international agreement on CFC controls. A lawsuit forcing the EPA to act unilaterally on regulations might threaten an international agreement, an EPA attorney explained to Miller. The NRDC obliged the agency for a short time, but when it became apparent that an international agreement to control CFCs would not be reached in time for the scheduled March 1985 meeting in Vienna, the NRDC filed the lawsuit.

The suit, filed in November 1984, sought to force the EPA to place a cap on overall CFC production as was mandated under its Phase Two proposals. The NRDC argued that, under the Clean Air Act, the EPA was required to regulate CFCs if they were deemed harmful to the environment. The group claimed the EPA had acknowledged this in its 1980 proposed regulations that had sat, collecting dust, for four years of the Reagan administration.

□   □   □

Besides the NRDC's lawsuit, the new EPA administrator could see that things were starting to change regarding the CFC-ozone issue.

International negotiations had been faltering since the late 1970s, but 1983 had marked a turn in the issue. In April 1983, Norway, Finland, and Sweden submitted a draft protocol for a worldwide ban on aerosols and controls on all uses of CFCs—a draft that became known as the Nordic Annex. Although the United States had led all countries by being the first to ban CFCs in aerosols and by proposing additional regulations (the Phase Two plans), the Nordic Annex represented a challenge to current U.S. policy and became a catalyst for change.

The science, too, was beginning to change. The first models adding other molecules to ozone depletion calculations had shown minor or no ozone losses. But most scientists knew the one-dimensional models had flaws and that better models might give radically different estimates. With studies showing that uses of CFCs were growing again, the data going into the computer models would have to be changed.

Ruckelshaus had his own thoughts on how the issue should be handled. In particular, he placed responsibility for stratospheric pollution matters on Joe Cannon, an associate administrator under Burford and the founder of the agency's greenhouse program. Cannon and John Topping, staff director of the EPA's Office of Air and Radiation, viewed CFCs as part of the greenhouse issue. But because CFCs were under the direction of the Office of Toxic Substances, Topping and Cannon had no legal mechanism to rule over the issue.

The Toxics Office had been given the responsibility for ozone protection dating back to the aerosol ban, although there were those who felt another office should be handling the matter. The Toxics Office had opposed any departure from Burford's conviction that a framework should be hammered out before a protocol, and a bitter fight developed between the Toxics Office and the International Activities Office throughout the summer of 1983. When it was finally decided that the ozone issue should be handled by the Office of Air and Radiation, the Toxics Office was glad to wash its hands of the matter.

In the EPA Office on International Activities, this change in authority was good news. Senior staff officer Jim Losey and a small group of colleagues had long felt that the agency's position on CFCs was wrong. Having won its battle within the agency, however, Losey sent a letter to the State Department in early September stating that the

EPA supported the Nordic Annex as it applied to a worldwide ban on aerosol uses.

"At the very least, we ought to be able to support what we've already got," Losey told State Department representatives.

The EPA's position surprised the State Department and irritated Mary Rose Hughes, the U.S. negotiator at the international meetings and a Burford ally. Hughes' embarrassment at EPA's switch was understandable. In prior international meetings, the United States' official position was that CFC controls might be necessary but a framework should be agreed to first to pave the way. But, Losey and others charged that, informally, Hughes was saying, " 'If we had to do it all over again we wouldn't even regulate aerosol uses.' "

While prodding the State Department to act on the Nordic Annex, EPA officials began trying to convince U.S. chemical manufacturers that getting other countries to adopt limitations on aerosols wouldn't cost them anything and would buy some insurance against ozone depletion. The manufacturers fought this suggestion, however. Many of the companies had plants overseas. But the EPA officials pointed out that the Clean Air Act had set a tough standard, and if the agency didn't pursue some type of regulations internationally it could be vulnerable to lawsuits.

While the industry balked, the EPA succeeded in making its plan the formal U.S. policy. After all, how could the State Department second-guess EPA on an environmental matter? Grudgingly, Hughes took the U.S. position to the next round of international meetings.

"It was the United States saying, 'Yes, there is a problem.' And we hadn't been saying that since 1979," Losey said. "So the four-year hiatus had come to an end."

The change in the U.S. position shocked the European Economic Community. The EEC was adamantly opposed to the Nordic suggestion of a worldwide aerosol ban. And now they had lost the United States as an ally on the issue. Without the Americans on their side, the Europeans were forced to make changes in their stand. And, despite repeated head-bashing in negotiations during the following year, the EEC eventually shifted from a "no protocol" position to a position that would set limits on CFC production—a stand they had already taken.

To Losey, the Europeans' position was disingenuous, although he felt the same might be said about the U.S. position. The Europeans' production cap had been set at a level it would take them decades to reach. But, he reasoned, at least they were admitting there was a problem.

The block of nations comprised of the United States, the Nordic countries, and Canada considered forcing a vote on the aerosol ban. But EPA officials discouraged this tactic. EPA officials reasoned that a worldwide ban on aerosols without the European community wouldn't have much effect on a global problem. And the State Department still didn't view the ozone issue as important enough to create a problem with a close ally.

Instead, the United States-Nordic block opted for a compromise. It offered to drop its push for an aerosol ban if all parties would agree to a resolution to the proposed Vienna Convention (it would be signed at a UNEP meeting in Vienna) calling for an agreement to resume negotiations on a protocol through UNEP by a specified date. And, prior to the resumption of negotiations, a series of workshops would be held to address the problem in a nonpolitical atmosphere.

By March 1985, a convention had been written and rewritten five times and finalized in six languages (only one of the minor details that must be attended to in international negotiations).

The Vienna Convention set up the basis for protecting the ozone layer through research, monitoring, and information exchanges with the understanding that more specific protocols or subtreaties to the convention would follow. The document called for all nations to assume responsibility for protecting the ozone layer but promised little more than improvements on the exchange of data during future negotiations. This was important, however, as several countries had previously refused to share information on CFC production within their boundaries.

Last-minute efforts were made to strengthen the convention with an actual agreement to begin reducing CFCs. Prior to the March 22 conference, UNEP's executive director Mostafa K. Tolba convened a two-day, informal meeting between negotiators in an attempt to add a specific protocol for CFC restrictions to the convention. But the pressure to resist further restrictions was heavy. Tolba's meeting was attended by the European Council of Chemical Manufacturers' Federations, the Federation of European Aerosol Associations, and the International Chamber of Commerce. Even the official delegates for Japan and West Germany were industry spokespersons. Conspicuously missing were any representatives from nongovernment organizations, such as environmental groups, which, some American negotiators thought, might have made a difference in signaling the United States' commitment to the issue.

The last-minute bargaining failed when the EEC refused to agree to any controls beyond what it already had in place. The participants, however, agreed to continue to work on drafting a protocol. And, of the 43 nations represented in Vienna, 20 signed the convention on the same day.

Tolba, in a statement made in Vienna, noted that the convention was significant simply because it dealt with the future instead of the selfish concerns of the present.

This convention, as I see it, is the essence of the anticipatory response so many environmental issues call for: to deal with the threat of the problem before we have to deal with the problem itself. As we all surely realize in the case of the ozone layer, facing the problem itself might already be too late. That we are taking the anticipatory approach is a sign, I think, of a political maturity that has developed over the years, which recognizes how vital it is that we act to prevent environmental degradation or disaster with wisdom and foresight.

We act now for the future. Those who could be threatened are the future generations that will have to live in a world that, through errors in judgment or mere short-sightedness, we risk making uninhabitable.

□   □   □

For Rowland and Molina, the early 1980s were deeply frustrating. While the Vienna Convention could be considered progress toward an eventual ban on chlorofluorocarbons, there was still a long way to go. Science had still not convinced the world's leaders of the gravity of the threat. And just how much scientific opinions matter in the upper echelons of decision making was open to question.

In general, American scientists are accustomed to working in a problem-solving manner, free of outside influences such as politics, economics, and value judgments. It is because of this perceived unbias and quest for truth that policy makers often rely on scientists for accurate assessments on issues of national and world importance. Up to a certain point, politicians defer to scientists.

But eventually, as in the case of CFCs, the issue becomes much more than a scientific one. At some point, scientific opinion becomes less important as other biases and considerations come into play. What evolves is a blend of science and politics, and scientists who choose to remain in the game from this point on are often viewed with suspicion.

The legitimacy of science can be questioned when scientists are asked to become more involved in political issues, wrote John S. Perry, staff director of the Board on Atmospheric Sciences and Climate of the National Research Council in a 1986 issue of *Environment*. "The unquestioned authority and legitimacy of any statement labeled 'science' has long since been eroded, if it ever existed."

Rowland, Molina, Cicerone, and a few others had to fend off questions regarding their credibility throughout the CFC-ozone debates. By calling for political action, they became targets for critics who suggested that scientists maintain a limited role in courtrooms. Even those who support the role of scientists in determining public policy are often reluctant to see that role magnified.

In his September 1976 speech at Utah State University, former White House science adviser Russell Peterson said:

> Inherent in virtually every environmental or public health policy decision are two components—scientific determination and social value judgments—and we as scientists should be careful to make the distinction. There are no individuals better qualified to make scientific judgments than scientists. But scientists are no more qualified than anyone else in making social value judgments. . . . The decision to regulate is the social value judgment. It answers the question, "Is the threat of risk worth the estimated economic dislocation?" This is not a scientific question but a value judgment.

And yet, scientists today are asked to take a more active role in policy making as issues of science and technology figure into major decisions regarding the nation's welfare. Scientists should feel obligated to state what they know to be true, advised Karl S. Pister, dean of engineering at the University of California, Berkeley, writing in the winter 1988 edition of *Issues in Science and Technology*.

"The science and engineering community must do a better job of helping our political representatives understand the scientific and technological ramifications of their decisions," Pister stated. "As scientists and engineers, we must articulate the problems as we see them."

The government, in fact, is in desperate need of more scientific advisers, claimed a panel of scientists speaking at the 1988 annual meeting of the American Association for the Advancement of Science in Boston. Policy makers should have free access to scientific opinion, and cooperation among science and industry must be encouraged without the fear of compromising research, panel members suggested. These goals could be met by appointing competent scientists to leadership positions in government agencies that deal with science issues, they said. Because scientists have traditionally viewed this blend of science and politics as inappropriate, however, few might be persuaded to accept such positions.

Too many scientists are unwilling to cross the line between scientific research and speculations on the implications of their research. Chemists joke about the First Commandment of Chemical Politics: Thou shalt not suggest anything that menaces anyone's profits, wrote the late Don L. Bunker, a University of California, Irvine, chemist, in a November 1974 issue of *The Nation*, during the first burst of the CFC-ozone depletion argument. "Too many scientists subscribe to a milder form of the same thing—don't speak up until you are absolutely sure of your ground. Thus we contribute to the presumption that any industrial procedure is innocent until proven guilty beyond any possibility of doubt."

Many of the scientists involved in the CFC-ozone debate were asked for their opinions on how the government should proceed regarding regulations. While most were willing to perform their public duty by testifying before lawmakers about their scientific research, many of the scientists were uncomfortable when the discussion turned toward solutions. The scientists were untrained and ill prepared to speak out on political matters and considered it inappropriate to answer questions outside of science. But they faced legislators who often couldn't understand why the stammering scientists were so hesitant to offer an informed opinion.

In one famous exchange during hearings before Senator Dale Bumpers' committee in 1976, Jim Anderson found himself in such an uncomfortable position. Bumpers' committee was a high-class affair, pursuing the CFC-ozone issue with thoroughness and integrity. And Bumpers took it seriously. When Anderson finished testifying about his research, Bumpers asked him pointedly whether government should ban chlorofluorocarbons.

Bumpers: "So I assume that you would perhaps, based on the evidence and the conclusions that have been reached so far, that. . . Congress ought to consider the banning of the unnecessary aerosols as quickly as possible."

Anderson: "Before I make a statement on that, I should point out that the measurements that I am involved in are crucial to the questions and I would like to remain neutral on such a question as you ask until I satisfy myself of the results of those measurements. That is from a scientific point of view. From the personal point of view, I feel very strongly about the issue of protecting the very delicate ozone photo-chemistry, and from that point of view I would urge on the basis of the data and calculations already available that action be taken."

Bumpers: "But you say that the conclusive results on chlorine on the stratosphere should be available within a year of process time. That is the reason I asked would you suggest that this committee perhaps wait until these measurements are available before suggesting regulatory bans of the aerosols?"

Anderson: "I watched very carefully the response of—"

Bumpers: "I am not trying to put you in a box of any kind. Just give me a yes or no answer."

Anderson: "I appreciate very much the statement. I do feel that the four walls are closing in."

As spectators roared with laughter over the exchange, Anderson advised the committee to take action to curb CFC use in more frivolous products, such as cosmetics, while allowing use of the chemicals in more needed products, such as refrigeration, for the time being.

Anderson paid a price for his statement, however. A few weeks later, he gave a scheduled speech at Du Pont and was questioned by a group of company salespeople who were angry over his comments to Bumpers—a confrontation Anderson found very embarrassing.

For Rowland and Molina, however, the decision to speak out on the ramifications of their findings was an intrinsic one. The two chemists never discussed the matter. They simply assumed that, in announcing their findings at the 1974 American Chemical Society meeting in Atlantic City, they would also call for regulations on the chemicals.

"I guess we might have had the choice of washing our hands,

like ivory tower scientists, and stepping back and saying, 'Well, here is the science, we'll let other people, environmental agencies, worry about it. We'll just do the clean science. But that never crossed our minds'," Molina said.

There is an attitude among many scientists, which is sort of snobbish, that doing that sort of thing, not just with environmental issues but with science in general. They say "I'm just going to do clean, pure science. If it has some kind of application whatsoever, I'm going to step back and do something else." I think that's very wrong. From the point of view of applications, one should always look to see whether there is an important consequence. After all, society is funding all of this. But it's still a very common attitude. Even more so with environmental issues, because there you have to take a stand. It's not just a matter of practical use but avoiding disastrous effects.

For Rowland, speaking out on the implications of research only made sense.

"What's the use of having developed a science well enough to make predictions if, in the end, all we're willing to do is stand around and wait for them to come true?" he told a *Newsday* reporter. He lamented that participants at scientific conferences often failed to discuss the ramifications of their studies. In a series of insightful interviews given to Paul Brodeur for the *New Yorker*, Rowland said, "I have never failed to wonder at how completely the sheer technical aspects of stratospheric science dominate such gatherings, and how little discussion, either formal or informal, is given to the implications of ozone depletion upon plants, crops, fish, weather, or for that matter, human health."

By suggesting an appropriate public-policy action, however, scientists run the risk of becoming known as crusaders who have compromised their scientific objectivity.

"He perhaps does, at times, take things to their extreme," one scientist commented of Rowland. "If there is a possibility there would be a bad effect but there is some uncertainty about it, he perhaps does have a tendency to minimize what he says about the uncertainty and maximize what he says about the danger. But he does it because he really cares and not for any other reason."

Because he was well established at age 46, when the CFC-ozone issue evolved, Rowland dismisses the idea that his involvement in the CFC-ozone issue might have affected his career. But there were times, he, Molina, and Cicerone have admitted, when the burden of speaking out grew enormously heavy.

While Molina says his career did not suffer because of his very public position on CFC regulations, the spotlight caused him a good deal of concern at times.

"I worried about my image as a scientist with other scientists in terms of going out on a limb and trying to make a case without appropriate background or foundation. That's why I work hard at getting appropriate background. I didn't just want to be an advocate without any substance behind it."

For Rowland, there was a higher price to pay. As Molina took on new responsibilities at NASA's Jet Propulsion Laboratory, it befell Rowland, who remained at Irvine, to continue testifying before policy makers on the status of the issue. Although Rowland was respected by his colleagues, few publicly supported his call for regulatory action. This was particularly true in the early years of the debate, prior to the 1976 NAS report which Rowland has always considered confirmation that his theory was correct.

"My feeling was that I was not quite alone. There were three of us: Mario and Ralph Cicerone and myself," Rowland said of those who supported CFC regulations. "You realize how alone you were, that basically, people recognize that (the theory is) controversial. They think it sounds good, and if it were not being talked about on the front page they would accept it. But since it's a hypothesis that called for political action they were prepared to wait for more testing."

According to his friend, Ralph Cicerone, it is difficult for any scientist to speak out without the support of his or her colleagues.

"There just weren't a lot of people picking up on this and being willing to talk to the press and to be assertive and say 'this is a very important global environmental problem,' " Cicerone said later, commenting on the minor damage the CFC-ozone debate did to his budding career while at the University of Michigan. "There just weren't any other scientists who were willing to do that. I think the cumulative effect of that was tough (on Rowland)."

Even after the 1976 NAS report confirmed that the chemistry of the theory was correct, Rowland's controversial stand about regulations affected his career for a period of several years. Until the mid-1980s, he received no invitations to speak at chemical industry meetings and few requests to speak before various university chemistry departments, as is customary for scientists of his stature. When he was asked to speak on college campuses, the invitation usually came from an academic department other than the chemistry department.

Rowland may have been shunned for another reason. During the late 1970s and 1980s, he expanded his efforts into field research as he and his team traveled around the world, collecting and analyzing air samples for trace gases. As important as such research may be, Rowland felt that many chemists looked down on such studies, considering only laboratory work to be "pure chemistry." Field work was considered by many chemists to be something else, perhaps geophysics or meteorology, but not chemistry. And the fact that Rowland's work was central to an environmental debate made many chemists uncomfortable.

"Chemists have become very defensive of all the talk about chemical pollution," Rowland said. "Chemists tend to be very paranoid about it."

A major highlight in his career, however, was Rowland's election, in 1978, to the National Academy of Sciences. It was an honor Rowland viewed as "approval from the scientific community for what you're doing. It doesn't get much involved with any comments you might have made with regard to regulations or what ought to be done. Rather, it's really a question about scientific contribution."

Among policy makers and the public, Rowland and his supporters might have suffered from a poor understanding that science works by a process of advances, corrections, mistakes, and new advances. This was something Rowland accepted. And although it may have affected his reputation, he succeeded, for the most part, in not taking much of the criticism directed at him personally.

"In science, you always argue," he said. "The way that one makes progress in science is by asking questions. That's something you're immersed in all the time. What you end up doing is looking for experiments that challenge the existing theories. Physical and chemical theories are proposed by people. And you realize fairly quickly that you are, in fact, challenging them."

But when the issue was as high profile as ozone, sometimes "corrections" in the normal scientific process were seen as scientists simply not knowing what they were doing. The 1976 chlorine nitrate controversy was a case in point. At the ozone conference at Utah State University, scientists held a rare, free-for-all discussion of their images and the credibility of science regarding the CFC-ozone depletion theory. Some participants suggested nervously that the chlorine nitrate finding had damaged their credibility. In response, Rowland described a class he taught at Irvine in which he gave every student a science article from a newspaper of 10 years ago that turned out to be wrong.

"It's no problem getting 50 articles that turn out 10 years later to be wrong," Rowland said. "I don't think the credibility of science is an absolute, inviolate thing. . . . There's no reason why we should defend the credibility of scientists who have made wrong statements."

To Rowland, the irony of the chlorine nitrate controversy was that—during arguments on how much less ozone depletion would occur—the very fact that ozone depletion would still occur was almost overlooked, proving that being essentially right is not always enough to protect one's reputation. "If you had more chlorine you were going to lose ozone. That's been part of the conclusion since the beginning. That was never really in doubt, although the fact that it was not in doubt was often obscured," he said.

As the years wore on, however, the fact that Rowland and Molina's basic theory had been deemed correct—that ozone depletion would occur due to the release of chlorofluorocarbons into the atmosphere—became of secondary importance. The arguments over how much depletion would occur and when overshadowed their basic findings. Did the prediction that eventual ozone depletion would be 5 to 9 percent instead of the 7 to 13 percent Rowland and Molina first suggested make them wrong? Sometimes, as industry and policy makers called for more data and more time to study the problem, it seemed that way.

□ □ □

By the summer of 1984, Rowland's disgust over 10 years of political and scientific waffling was evident. In an interview with the *New Yorker* he said, "From what I've seen over the past 10 years, nothing will be done about this problem until there is further evidence that a significant loss of ozone has occurred. Unfortunately, this means that if there is a disaster in the making in the stratosphere we are probably not going to avoid it."

It would give Rowland no satisfaction to learn, a few months later, that he was right. A disaster was in the making. Half a world away, at a British outpost in Antarctica, Joe Farman and his team of workers with the British Antarctic Survey were trying to figure out why their ground-based monitoring stations were showing a 40 percent drop in ozone.

8
The Ozone Hole
October 1984—March 1986

A sense of excitement filled Joe Farman as he hurried to his Cambridge office. He could scarcely believe what he had just heard at a meeting of his scientific advisory group. But the figures on the sheet of paper he now held bore the whole thing out. Those curiously low ozone values his research team had been detecting each spring over Antarctica had held up again this year, 1984. The sky really was disappearing over the South Pole. It was, as Farman was to repeat again and again, "dreadful."

For almost three years Farman had kept mum on this mysterious secret, even quashing the enthusiastic attempts of a young Cambridge chemistry student who had made the Antarctic ozone readings part of his Ph.D. thesis and wanted to publish some of the results long ago. Farman, a tall, thin man who speaks rapidly and is most comfortable in plaid flannel shirts, was nothing if not cautious. In 1982, when the readings in Antarctica began showing a dramatic dip in ozone, Farman was more than a little skeptical. If they were wrong in their interpretation of the data, Farman knew what little support he had from his government would vanish. And visions of his precious research group becoming the laughingstock of the international community of atmospheric scientists was too much to bear. Better to move very carefully.

Farman was proud of his small band of researchers called the British Antarctic Survey, although the group was relatively unknown to anyone outside England. Since 1957, Farman had been dispatching research assistants to the British outpost at Halley Bay, Antarctica, halfway around the world, to measure various trace gases, such as CFCs, and total ozone. For 25 years, the research efforts hadn't yielded anything of great importance, but it was Farman's longevity in the Antarctic that helped him spot low ozone readings in 1982.

"In a sense, we were in a position to be able to tell people that things were changing because we had a good, long record," Farman said later, perhaps justifying the fact that a virtually unknown research group had made one of the most important atmospheric discoveries of the decade.

The British readings were made using the ground-based Dobson spectrophotometers. Developed by Sir G.M.B. Dobson, the oldest operational unit had been put in place in 1931 to help meteorologists track air movements and measure ozone by recording how much ultraviolet light reaches the earth. Since ozone absorbs UV light, less ozone would mean more UV light reaching earth.

The Dobsons in Antarctica were notoriously difficult to keep working, however, and the British equipment was old. So when the first record of low ozone values came in after the 1982 spring season, Farman immediately suspected equipment failure. What's more, ozone values were known to fluctuate naturally and the American satellites hadn't spotted any ozone depletion in the area. Farman viewed the readings with skepticism and ordered new equipment to be sent to Antarctica for the next year's measurements.

A closer inspection of the data revealed, however, that ozone values had actually begun to slip as early as 1977. And when the 1983 data also showed a decline, Farman decided to try a different measurement to verify the ozone loss. Dobson instruments measure ozone from the ground to one point directly overhead. Perhaps there was something about the Halley Bay station that was unusual. During the 1984 season, Farman turned his attention to a second measuring station over Argentine Island, 1000 miles to the northwest. When the 1984 readings were completed, the British group found itself looking at a huge 40 percent ozone loss during a period of about 30 days, from September to October, over both Halley Bay and Argentine Island.

According to the British observations, the plunge in ozone values occurred at the end of each austral winter, just before the polar vortex broke up and, apparently, filled the "hole" in with ozone-rich air from elsewhere in the hemisphere.

The polar vortex, which forms each year after the March equinox, marks the onset of the southern winter. The vortex itself is a swirling mass of very cold, stagnant air surrounded by strong westerly winds.

It is an extreme weather phenomenon, formed because of the unique geographical features of the Antarctic which was surrounded by oceans and devoid of the mountains and land masses that might lead to more varied circulation patterns. There is a polar vortex in the Arctic, too, although it is not as strong. In previous years, Farman's Antarctic readings had shown ozone values in the vortex each winter and spring to be around 300 Dobson units, increasing to about 400 Dobson units in the early summer when the vortex breaks up. But Farman and his group were observing readings of only 140 Dobson units.

Now, in October 1984, Farman sat with colleagues Brian Gardiner and Bob Murgatroyd in a meeting room at Cambridge, stared at the latest data, and went down their checklist one more time. They had checked the instruments. They had checked their calculations. What else could possibly be wrong?

"This is it," Farman told the group. "We must say something. It's perfectly obvious something strange is happening. We have to get this out as quickly as possible."

Heading back to his office to begin writing up the group's paper, Farman couldn't help but wonder how his colleagues would react to the publication of the Antarctic findings. He wondered who would believe him. Easy going and somewhat self-deprecating, Farman was, nevertheless, a little frustrated that his British Antarctic Survey had been ignored for almost 25 years. It was true that ozone measurements were difficult to do. There were Dobson stations around the world that were considered poorly run and that no one regarded as accurate. Every Dobson station had to be questioned. But, after all, hadn't he been at this longer than practically anyone?

His superiors' reactions would be another story. For years, various government and academic leaders had tried to stop his survey. On many occasions, Farman had sat and fumed while the British House of Commons doled out a paltry 400,000 pounds a year for stratospheric research of which he regularly took his "trivial" portion: about $18,000.

"You've been doing this for 30 years. Why don't you stop?" people said to Farman on various occasions. Sometimes, the battle to keep his small program growing seemed on the verge of defeat. Early in 1984, the director of the Natural Environment Research Council, Farman's employer, came around to review what the group was working on. Enthusiastically, Farman let the man in on his secret, showing him the latest data on Antarctic ozone levels and explaining how they were also measuring chlorofluorocarbons. The man eyed Farman quizzically.

"What are you measuring them for?" he asked.

"Well, quite simply, there is a big CFC industry and people are writing that ozone will change. And the only way you can tell if ozone has changed is to sit and keep measuring it," Farman said.

"Oh," Farman's superior mused slowly. "You're making these measurements for posterity! Well, tell me, what's posterity done for you?"

Blanching, Farman chose to drop the matter. Later, he called the remark "a typical Cambridge, high-tech remark. It was dreadful."

The simple fact of the matter was that, although Sherry Rowland and other major atmospheric researchers hadn't heard of Joe Farman, he knew of them and their work. Farman knew that the Rowland-Molina theory predicted declining ozone values due to CFC emissions. What was so confusing, Farman thought, as he sat down to write up the British findings, was to see a rapid, huge, seasonal variation instead of gradual change.

"We thought, gosh, this really must be the chlorine at last, and it's appearing in a way in which we weren't expecting," Farman later recalled. Working out the details of their paper, Farman, Gardiner, and J.D. Shanklin concluded that the bitter cold Antarctic temperatures and the presence of CFCs caused an accelerated ozone loss each spring. To Farman, the whole problem looked nightmarish.

"It suddenly became very clear how it could happen. What wasn't so clear was how in the hell it could ever stop," he said. "I'm still not sure I understand that."

□   □   □

Farman's paper arrived in the offices of *Nature* magazine on Christmas Eve, 1984. Within weeks, a copy was sent to Susan Solomon, a 29-year-old atmospheric chemist at the National Oceanic and Atmospheric Administration in Boulder. Solomon had been asked to referee the paper, to help judge whether it was suitable for publication in *Nature*. Organized, sensible, and not easily ruffled, Solomon read the paper in a state of shock. The brief summary before her, written by a group of people she'd never heard of, was about to change her life.

Although she was young, it was not surprising that Solomon was doing things like refereeing papers for *Nature*. By all accounts, she was a brilliant, aggressive scientist who did high-quality research. Solomon, who has dark hair and the kind of skin that rarely sees the sun, had had the best of training. She had earned her Ph.D. under Hal Johnston at Berkeley and completed her thesis at the National Center for Atmospheric Research working for Paul Crutzen. She joined NOAA after earning her doctorate in 1981.

It was at undergraduate school at the Illinois Institute of Technology, however, that Solomon fell in love with atmospheric science. In the lab where she was working, some researchers had been measuring rate constants for gases that turned out to be important in the atmosphere of Jupiter. "I just thought that was the neatest thing I'd ever seen," Solomon said. "It was an incredibly exciting application of chemistry in the real world rather than in a test tube." She was hooked.

Now, reading Farman's paper, Solomon was quickly convinced that the work was legitimate. It seemed to be carefully done. She couldn't find anything wrong with it. And it was obvious that, if the British researchers were right, they had found something extremely important. Solomon returned the paper to *Nature* with the suggestion that it be published.

Not everyone was as impressed as Solomon, however. The Farman paper appeared on May 16, 1985, and there was immediate criticism of the paper. It seemed ludicrous. "Who in the hell is Joe Farman?" noted one researcher. Even the scientists who had studied ozone depletion since the early 1970s were stunned.

Ralph Cicerone had left Michigan for a position at the Scripps Institute of Oceanography in San Diego and then moved to the National Center for Atmospheric Research in Boulder. He thought the paper could be important but understood the widespread American skepticism.

"The British Antarctic Survey is not a household word," he said. "At the time, most of us had never heard of it, had no idea whether these people did good work. You couldn't automatically give credence to the work."

And Cicerone could see other problems. Although the British researchers had confirmed their Halley Bay reading with a second measurement over Argentine Island, the paper was based on measurements taken at two isolated points. If huge losses of ozone occurred over Antarctica each spring, why hadn't the satellites picked it up?

At the Jet Propulsion Laboratory, Mario Molina read the *Nature* article and suddenly felt like he had been shut out of his own field. Normally, scientists hear about any important papers in their field long

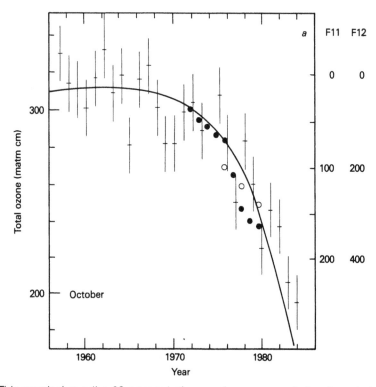

This graph shows the 30 percent plunge of ozone over Antarctica during October 1984 as detected by British scientist Joe Farman. Ozone fell from its normal range of about 300 Dobson units to under 200 Dobson units that month.

before they are published. By the time the research is published, it's usually old news. Why hadn't he heard about this? Molina wondered. Did everyone else know about it? But Molina was excited. "It must be chlorine, and we must not have considered the very different conditions in Antarctica," he told colleagues at JPL.

At Harvard, Mike McElroy also found out about the Antarctic depletion by reading the *Nature* paper. He, too, suspected that some sort of unusual chemistry was taking place in the unique Antarctic environment. But, McElroy thought, whatever was happening, it must be very different from any mechanism or model scientists had proposed in the past. This ozone loss was phenomenal.

Rowland was as perplexed as anyone. He had never heard of Farman. But the paper was carefully crafted and the British researchers had made a convincing argument that the values were not due to their instruments going awry. Anxious to find out what others thought of the paper, Rowland departed for Switzerland where, in July, about 40 leading atmospheric scientists were scheduled to meet to review a NASA publication summarizing the state of the stratosphere. At the meeting, both Rowland and Molina raised the subject of the Farman paper. Both were somewhat surprised to find that people weren't terribly interested in it.

"What do they have in the way of confirmation?" the scientists asked when Rowland mentioned the British paper. "If there really was a big hole there it would have shown up in the satellite measurements."

There simply wasn't any confirmation at that point, although there was one tiny bit of support for the paper. Unfortunately for Farman, the one piece of evidence in his favor came from someone even less well known.

Earlier in 1984, a Japanese researcher named Shigeru Chubachi had published the results of a ground-based ozone measuring station at the Japanese Antarctic outpost in Syowa. Chubachi had also seen low ozone levels, but did not have enough data to see the trend that Farman had detected over many years. This was perhaps due to the fact that the British station was farther south while Chubachi's station at Syowa may have been on the edge of the vortex and, therefore, not as likely to pick up on ozone losses. Chubachi, who had the added problem of not being fluent in English, had presented his findings at an ozone meeting in Greece in 1984. Later, Rowland, who takes copious notes at such meetings, found he had lost his notes from the Greece meeting. But, he said, "There had been reports from time to time of low ozone values over Antarctica. There was certainly no discussion at that meeting in Greece of unusual problems over Antarctica."

With a Xerox copy of the British paper in his briefcase, Rowland left Switzerland for Hawaii where he met Brian Gardiner, one of the co-authors of the British paper. Rowland was impressed with Gardiner and became completely convinced of the validity of the British research. His response was something of a relief to Gardiner. Gardiner could see that at least one member of the atmospheric community was interested in the findings.

Someone else was highly interested in the British findings. The paper deeply troubled Donald Heath's research group at NASA's Goddard Space Flight Center. Heath was in charge of reviewing the satellite data that might have detected a large ozone loss. With trepidation, the Goddard scientists realized they had overlooked the computer data that showed the ozone hole. The instruments sometimes gave measurements that were known to be anomalies, and the Goddard scientists had programmed their computers to throw out any measurements of ozone below 180 Dobson units, which was considered impossible. Another problem was the time delay in reviewing the satellite data. Because the satellites accumulate so much information, months can elapse between the time the measurements are taken and when they are read.

Heath claimed later that his group had spotted the unusual numbers several months before the British paper was published and were simply trying to figure out, in the secrecy of their offices, whether they were real. He was confident in the instrument but thought there might be a problem with the telemetry. The scientists relied on tapes to tell them the exact location of the satellite and the time of day. In the past, they had had problems with some of the tapes telling them exactly where the satellite was looking. If the location was wrong, Heath reasoned, he would calculate the wrong angle of the sun illuminating the atmosphere which would give him an incorrect reading.

In any event, the satellite and computers were telling the truth, and the Goddard group had missed the chance to make one of the most important scientific discoveries of the decade.

"We saw it in the satellite data, but it was so large and so unusual that we really had serious questions about whether it was real or not," Heath recounted. "We were more or less in the process of saying, 'Look, do we believe it or don't we believe it.' "

Farman's paper quickly convinced Heath that the satellite data was correct. Frantic to figure out just how much depletion might be occurring, the Goddard team dug up reams of computer printouts showing total ozone values from the past several years. The Nimbus 7 satellite took ozone readings from 600 miles above the earth's surface. Pouring over the data, the Antarctic ozone hole suddenly appeared before them.

The Nimbus 7 showed an ozone hole the size of the continental United States centered over Antarctica with ozone levels down to 150 Dobson units.

Within a month, Goddard scientists had corrected their computers and, at a small meeting in Austria in August, Heath showed colorful slides of the ozone hole from 1979 to 1983 that left the audience stunned. The British researchers had their confirmation. But, poor Joe Farman; instead of congratulations, people were now demanding why the group hadn't published their findings sooner.

In his good-natured way, Farman laughed off the criticism. "I sometimes feel we should have (acted sooner)," he told his critics. "On the other hand, the very fact that we delayed it until it was absolutely certain meant that there was never an argument. It was accepted. That's the real trouble with all these environmental problems. Too many people make too many noises all the time. Whereas if you only show it when you've got something to show then, OK, people understand it."

It didn't bother Farman that it had taken several months, until Heath's data came out, for people to accept the British work. But, while American scientists now scurried to determine what was causing the ozone hole, Farman was dismayed that the discovery had made little impression in his own country. Even when NASA's Bob Watson brought the Nimbus 7 data to Europe to see what the European reaction was, Farman could have warned him to stay home. "There wasn't much European reaction," he said.

In his own country, Farman now faced the wrath of the Imperial Chemical Industries, the industrial organization of British CFC manufacturers, who demanded to know why he had associated the ozone drop with increased CFC emissions. Farman defended himself by saying that he had chosen the explanation that he thought was most likely to blame. Not surprisingly, his requests for more research money for new equipment to confirm his work continued to be denied. If you ask Joe Farman what he's doing now, he'll likely tell you that he is still trying to raise some research money to buy new instruments.

                              □   □   □

The discovery of the Antarctic ozone hole baffled atmospheric scientists in every corner of the world. In the fall of 1985, there was still much to be explained. How was it that the computer models that had been overhauled and restructured repeatedly over the past 10 years had failed to predict a dramatic drop in ozone?

The failure to predict the Antarctic hole also raised questions about ozone predictions for the entire world. There was no reliable method for mapping ozone. The two techniques for measuring ozone—the satellites and the ground-based Dobson measuring devices—did not agree. According to Heath's data from the Nimbus 7, total global ozone had fallen 3 percent from 1978 to 1984, but an analysis of Dobson stations around the world had showed no total ozone changes in the same time period. The Dobsons, however, had only been analyzed up to 1980.

One problem with the satellite measurement was that the optical surfaces of the instrument were known to degrade over time. Critics contended that an observed change could simply mean the instrument was failing. Nevertheless, Heath's observations that ozone losses were most prominent in the spring and fall and in the upper northern and southern latitudes tended to match what computer models predicted, and thus merited further consideration.

The Dobsons had their own problems. There were about 40 Dobson stations in operation, mostly in the Northern Hemisphere. But various countries had their own standards for maintenance and varied on how frequently the instruments were calibrated, making a worldwide comparison of Dobson data difficult. Detecting global ozone changes by Dobson was also difficult because scientists interpreting the measurements had to take into account how ozone varies naturally from season to season, at different latitudes, and how ozone fluctuates with the natural solar cycle. For several years, studies of Dobson stations (many sponsored by the CFC industry or the Chemical Manufacturers Association) had shown no change in total ozone, and industry officials gave great credence to the long history of the Dobson network.

"Some of the finest statisticians in the world did these trend analyses," an official for the Chemical Manufacturers Association told *Chemical and Engineering News.* "When you do that for total column ozone, you see no statistically significant change in natural behaviors."

In fact, the Dobson readings had bothered Rowland for some time. Convinced that the Dobson analyses were not up to date and were not undergoing the proper scientific review (by 1984, of course, no one other than Rowland was too worried about ozone depletion anyway), Rowland asked a member of his research team, Neil Harris, to review data from a very old and reliable Dobson station in Arosa, Switzerland.

In 1984, scientists at the Swiss station had reported that their 1983 measurement was 8 percent lower than the annual average over the last half century. It was the lowest value ever recorded at the station. In addition, researchers in West Germany and Canada had found similar losses in 1983. Based on these findings, NOAA researchers concluded that, during the first half of 1983, ozone had dropped between 5 and 7 percent in the Northern Hemisphere.

But why were the statistical analyses of the Dobsons showing no global ozone loss? To Rowland and Harris, the discrepancy appeared to have something to do with seasonal ozone loss. The models being used, Rowland knew, assumed that whatever was going on in the atmosphere was the same during the summer and winter. James Angell of NOAA had analyzed the data by season and had seen some variation, and there had been suggestions that different statistical models, especially those that weighted the summer months more heavily, could be inaccurate. In reviewing the Arosa data by season, Harris and Rowland discovered that there was an obvious ozone loss in the winter but no statistically significant loss in the summer. The Irvine researchers then looked at Dobson stations in Bismarck, North Dakota, and Caribou, Maine, and saw the same pattern: ozone losses in the winter and little change in the summer. To Rowland, the message was clear: there was a winter-summer difference in ozone changes—less ozone in the winter and not as much change in the summer. This interpretation was later to prove crucial in understanding the 1987 finding of large ozone losses in the upper latitudes of the Northern Hemisphere during winter months.

The large ozone losses reported in Switzerland, West Germany, and Canada raised other questions, however. Several researchers had found it odd that the ozone losses in these areas had occurred in the lower atmosphere—12 to 20 miles in altitude—where most models said that ozone should be increasing. These calculations of increasing ozone in the lower stratosphere had resulted in the small ozone depletion estimates featured in the 1984 NAS report.

One possible explanation for the 1983 ozone loss was the eruption of the Mexican volcano El Chichon in April 1982. The volcano had spewed massive amounts of sulfuric acid particles and debris into the atmosphere. Perhaps, Rowland thought, chlorine nitrate (that all-but-forgotten chemical that Rowland and Molina discovered, in 1976, could alter ozone depletion mechanisms drastically) and other molecules might react on the surfaces of volcanic debris, leading to unusually rapid ozone losses.

This idea was an unusual one. In the previous 15 years of work on ozone depletion, scientists had dealt strictly with homogeneous reactions, reactions that occurred between uniform substances such as gases. No heterogeneous reactions—reactions between two nonuniform substances—had been shown to be important in the stratosphere. In late 1983, after El Chichon but before the Antarctic hole was discovered, Rowland had begun questioning the calculations of the models. He suggested that what might be happening in the atmosphere was a heterogeneous reaction such as between a gas and water. Perhaps some heterogeneous reactions were missing from the models, he thought.

After discussing the idea with Don Wuebbles, a scientist at the Lawrence Livermore Laboratory, and debating what other molecules might be involved, Rowland directed members of his research group to begin experiments with chlorine nitrate, hydrogen chloride (the chlorinated compound that forms in the atmosphere after the stratospheric decomposition of CFCs), and water. Hydrogen chloride and chlorine nitrate were important because both could serve as reservoirs for chlorine. Reservoirs, in this case, are compounds that do not directly deplete ozone but can, under certain conditions, release chlorine that can destroy ozone. Rowland's group found that chlorine nitrate did react with water and hydrogen chloride on laboratory surfaces. The researchers concluded that such reactions might also occur on the surfaces of particles in the atmosphere, such as dust thrown by volcanoes or ice particles in clouds, creating a much more rapid mechanism for chlorine to attack ozone when the chlorine is released from the reservoir molecules. The new values were inserted in Wuebbles and colleague Peter Connell's computer model. Without the reactions, the models predicted only small changes in total ozone over the next few years—the 2 to 4 percent predicted in the NAS report. But with the new data, the model showed ozone losses in the range of 20 to 30 percent. These calculations, which were completed in 1984, were reported in June 1984 at a meeting at Feldafing, Germany.

While the model was far from precise and very little was known or understood about heterogeneous reactions, by the summer of 1984 Rowland felt the conclusion of "no total ozone change" was suspect.

Molina, too, was exploring the idea of heterogeneous reactions. While working together at Irvine, the two had discussed and worried about the possibility of heterogeneous reactions in the stratosphere. Somewhat serendipitously, Molina, too, had been studying heterogeneous reactions involving chlorine nitrate and hydrogen chloride in 1984. He published a paper on his work in 1985 in the *Journal of Physical Chemistry* discussing heterogeneous reactions. While Molina, Rowland, Susan Solomon, and a few others were the first to jump on the idea that heterogeneous reactions were involved in the weird Antarctic phenomenon, it was Molina who eventually provided the complete explanation for the rapid ozone losses there.

□ □ □

Susan Solomon could not stop thinking about Farman's paper, a some-what odd change of heart for her because, for several years, Solomon had been uninterested in the problem of chlorine increasing in the atmosphere.

The subject of chlorine in the upper atmosphere was relatively straightforward. But the complicated issue of chlorine at low altitudes, with all those compensating factors that made ozone depletion so difficult to estimate, bored her. Instead of tackling the chlorine issue with traditional models, Solomon had, with colleague Rolando Garcia, become involved in two-dimensional modeling and the study of the dynamics of air motion and chemistry. Rowland, however, didn't let the young chemist off the hook so easily. The two had several discussions on why Solomon hadn't taken up studying the problem. On one occasion, at a June 1984 scientific conference in Germany, Rowland, his wife, and Solomon were engaged in a conversation when Joan Rowland said to Solomon, "Chlorine in the atmosphere is really very important. Somebody as bright as you are ought to be working on it."

Now, with the intriguing British paper, Solomon felt herself being persuaded to join in the chlorine chase. She felt that there was almost no chance that the dynamical motions of air could explain the Antarctic loss and strongly suspected chemistry.

Solomon recalled that a group of New Zealand researchers had published a paper in 1984 showing that nitrogen dioxide was very low in Antarctica. This was another clue that unusual chemistry might be taking place in the atmosphere because scientists knew that nitrogen dioxide, if present, would tie up chlorine and prevent ozone depletion. Solomon wondered if, perhaps, some American bias toward research done by foreign chemists had cost U.S. scientists some time in reacting to the situation in Antarctica. People were slow to accept Farman's work. And the New Zealand paper was not given serious consideration. "It's like we have to go measure these ourselves before we believe it," Solomon said. "I think if you look at what they published it was clear from the first that they made a good measurement; that, in fact, the (nitrogen dioxide) abundances in Antarctica were much lower than they were in the Arctic."

By the fall of 1985, Solomon had taken the Antarctica data and had tried to write a computer model that would explain how the huge loss of ozone had occurred. Like others who were also trying to model the phenomenon, Solomon used all the chemical reactions that were known but failed to get the computer to duplicate anything close to what had been observed. Then she began to think about the rare pinkish clouds, called polar stratospheric clouds (PSCs), that form over Antarctica each winter.

Together, Solomon and Garcia pieced together a two-dimensional model showing how polar stratospheric clouds in Antarctica, together with a coupling of hydrogen chloride and chlorine nitrate— the same combination Wuebbles and Rowland had studied earlier— might be the cause of the drastic ozone loss each October. She wrote to Rowland with the preliminary results later that fall.

As usual, Rowland was struck by the confidence of the young chemist. And he was pleased to see that she now thought chlorine in the atmosphere was an important enough matter to pursue.

"She said to me, 'This is very important,' and that it was very important that she should work on it," Rowland recalled. "That's confident. But given that Susan is brilliant, then it is not over-confident. And the solution to the problem was greatly aided by the fact that she was involved."

□   □   □

Joining forces, Solomon, Garcia, Wuebbles, and Rowland put together their explanation of what was happening in Antarctica and submitted the paper to *Nature*. The theory concluded that the cold and darkness of the Antarctic winter created unusual conditions for rapid ozone loss. The chemists suggested that harsh conditions in the Antarctic might somehow alter chlorine reservoirs, such as hydrogen chloride and chlorine nitrate, allowing the reservoirs to give up active chlorine that could destroy ozone.

They knew, for example, that most ozone in Antarctica could be found at 12 to 20 kilometers where, normally, chlorine is inactivated by being tied up in reservoirs. But, the scientists concluded that PSCs could provide a surface for heterogeneous reactions that free chlorine from the reservoirs. Polar stratospheric clouds are high-altitude clouds that form in the polar darkness when temperatures reach −120°F. Trapped in these ice crystals, compounds such as nitric acid could condense and freeze, becoming unavailable to react with chlorine.

The perplexing part of the theory was the explanation of how the chemicals could be converted from an inactive state to an active state so rapidly. The group suggested that chemical reactions simply take place much faster on ice surfaces than they do in the purely gaseous state. Possibly, the ice particles modify the chemical reactions, setting up a volatile laboratory beaker of compounds that suddenly react once they're hit with the first rays of sunlight in the austral spring. Rowland's lab studies showed that these reactions occurred on laboratory surfaces, such as glass and Teflon, and it was not implausible that they could also happen on icy clouds. But heterogeneous reactions were simply not well understood.

Solomon, Garcia, Wuebbles, and Rowland were not the only people who suspected a chemical process was at work. At Harvard, Michael McElroy, Steven Wofsy, Ross Salawitch, and Jennifer Logan suggested a similar set of chemical reactions and the involvement of PSCs. Both groups' papers appeared in print in *Nature* on June 19, 1986.

McElroy disagreed with the Solomon group's theory, suggesting that chlorine alone could not be responsible. While Molina's own theory was not published until the next year, he suspected that these reactions were not adequate and had begun working on a mechanism in which chlorine oxide reacts with itself to produce a chlorine oxide dimer. A dimer is two molecules of the same type reacting with each other. Molina suspected that the chlorine dimer provided the catalytic cycle of ozone destruction with Antarctica because it explained how one could recycle chlorine over and over again with great speed. While the other theories also recycled chlorine, Molina's explained the rapid recycling that was the key to the Antarctic loss.

As had been the custom since the early years of the chlorine-ozone debate, vehement disagreement on what was causing the ozone hole was to follow. During the fall of 1985 theories to explain the ozone hole were popping up, "almost as many as there are workers in the field," *Science* magazine noted.

In what became known as the dynamicists' theory, several respected atmospheric scientists suggested that air movement was the cause of the hole. The assumption certainly made sense. After all, the atmosphere is not a static layer of gases. Air masses are in a constant state of flux, causing a shifting of ozone and chemicals throughout the atmosphere. Normally, ozone is formed at high altitudes over the equator and is pushed to the poles and squeezed downward in the spring when the polar vortex breaks up. But, the dynamics theories (there were several focusing on air movement) suggested that some processes had somehow weakened the downward draft of this ozone-rich air.

A third theory was also receiving considerable attention. Called the "odd nitrogen" theory, it suggested that the solar cycle was to blame for the abnormal ozone fluctuation.

According to the major proponents of the theory, Linwood B. Callis of the NASA Langley Research Center in Hampton, Virginia, and Murali Natarajan of SASC Technologies, also of Hampton, ozone destruction was being catalyzed by abnormally high levels of odd nitrogen. Odd nitrogen is produced by blasts of particles from the sun. The solar particles—protons and electrons—tend to penetrate at the poles where the resistance of the earth's magnetic field is weakest. During the polar night, the odd nitrogen sifts down into the stratosphere where it is available to help deplete ozone. Callis suggested that the Antarctic depletion could be linked to the second largest solar maximum in 250 years, which occurred in 1979. According to this theory, an ozone hole would occur with each solar cycle, or every 11 years. But, the good news was that the Antarctic atmosphere would recover during other years.

To some observers, the solar cycle—or odd nitrogen—theory was favored simply because it explained the phenomenon in the simplest terms and meant that the hole was a brief aberration, not the first sign of impending doom for planet earth, as the chemical theories implied.

"It should turn around, if what we're suggesting is correct," Callis said at a spring 1986 meeting of the American Geophysical Union. And, noted one observer, "It's kind of nice because it's a natural solution."

Supporting the chemical theories, however, meant at least partial acceptance that humans were probably to blame for this huge smear on planet earth and that something had better be done to stop it—fast.

"It's like rolling dice," McElroy told *The New York Times*. "The big money question is if what's happening in Antarctica is likely to be a foretaste of what might happen in the northern region."

While each of the theorists pushed his or her own explanation, the chemists and dynamicists became polarized within months after the Farman data hit the grapevine. And each group seemed particularly bent on ruling out each other's theories.

"So far, there's a lot of wishful hand-waving," Jerry Mahlman, a

NOAA scientist and a supporter of the dynamics theory, told a *New York Times* reporter in criticizing the chlorine chemistry theory. Mahlman admitted that he had bet a Chinese lunch that ozone levels would rise again when the 1986 data came in. Still, he and everyone involved knew that more was at stake than fortune cookies.

9

The Sky Is Falling

March 1986—June 1986

Even people who understood little about science or the environment could see that the Antarctic ozone hole was a frightening phenomenon. NASA's satellite image—the swirling dark hole surrounded by bands of color—was shown repeatedly on evening newscasts. The image was vivid and the message was, too. The idea that the ozone hole, like a tiny puncture in a fragile eggshell, could signal the eventual demise of the earth's atmosphere shook many policy makers into action by early 1986.

Lawmakers were also unsettled by the release of the NASA-UNEP report on the atmosphere that had been reviewed in Switzerland in July 1985 (the meeting at which Rowland had raised the issue of Farman's paper). Coming on the heels of the Vienna Convention and the discovery of the ozone hole, the NASA report couldn't have been timed to capture more attention.

The report, an international effort by 150 scientists from 11 countries, warned that damage to the atmosphere was already apparent but that scientists knew too little to predict what devastation might lie ahead.

"What was once mainly based on theoretical predictions is now being confirmed by observations," the report concluded.

In a 2000-page text, the report stated that symptoms of the greenhouse effect were becoming apparent and that there was "compelling evidence" that trace gases, such as CFCs, were increasing on a global basis. If CFC emissions continue at 1980 rates, the report concluded, the average amount of ozone would fall 4.9 to 9.4 percent by the middle of the next century. (While the 1984 National Academy of Sciences report had put future ozone depletion at 2 to 4 percent, NASA's estimate showed the figures once again approaching the 7 to 13 percent range that Rowland and Molina had originally suggested.) The report mentioned the Antarctic phenomenon cautiously by saying that scientists lacked a proven explanation for the depletion. But it raised the question of whether the hole could be "an early warning of future changes in global ozone." In a pointed message to world leaders, the report suggested: "We should recognize that we are conducting one giant experiment on a global scale by increasing concentrations of trace gases in the atmosphere without knowing the environmental consequences."

Quite suddenly, the ozone issue of the mid-1970s was back and seemed more intense than ever. And Rowland and Molina's allies were quick to point out that the only thing that had really changed in the past 10 years was government's attitude about the problem. The problem had never disappeared.

"The No. 1 message of the report is that the original theory of how humans can alter the atmosphere globally is still hanging together," said Cicerone. "We have not been able to shoot down the theory that chlorofluorocarbons are destroying ozone."

The Antarctica hole seemed to revive Rowland's commitment to seeing CFCs banned. For eight years he had continued his basic research, quietly and tirelessly, carting small canisters around the world to obtain air samples and measure the samples for pollutants. (He once made a stop in Texas to test the air outside the Du Pont Company's huge CFC plant.) But the new attention to the theory also reopened old wounds. Sitting in his Irvine office one summer day, Rowland seemed annoyed that the effort he, Molina, and Cicerone had put into getting the CFC ban on aerosols had not amounted to anything more. A 1986 Rand Corporation report revealed that CFC emissions had risen to their preban levels. CFCs used in auto air conditioners, insulation, and fast-food packaging were growing fast. But even now, facing a lawsuit by the Natural Resources Defense Council, the EPA had not acted on those almost-forgotten Phase Two regulations to reduce other CFC uses that had been proposed during the Carter administration.

"The interest dropped off with the thinking that we've solved the aerosol problem with Phase One (regulations) and we'd work on Phase Two," said Rowland. "The problem is Phase Two never arrived and most countries have never gotten to Phase One."

Even if the Antarctic ozone hole were linked to CFCs, regulations of nonaerosol uses of CFCs would be much harder to come by, he warned. In the 1970s, consumers could protest use of the chemicals by refusing to buy deodorants, hair sprays, and cleaners containing CFC propellants. But in 1986, CFCs were much more deeply ingrained in American consumerism.

"In the '70s, the individual consumer had a vote," Rowland said. "You don't have that now because you can't go to an automobile dealer and say I want an automobile with an air conditioner that doesn't use fluorocarbon-12." Besides, he shrugged, this was not the 1970s. Consumers may simply not care if there is any ozone left to protect their grandchildren's tender skin.

"My own somewhat cynical view is that, on average, people in the United States don't worry about what will happen to *their* grandchildren," said Rowland, sighing. "They worry about what will happen to *their* children somewhat, but they expect their children to worry about what will happen to their children. And as far as industry is concerned, they have great difficulty looking more than 10 years down the road."

With newspaper and magazine reporters from around the world once again calling his office to ask about the curious ozone hole, Rowland repeated his challenge to the CFC industry.

"Industry always said that we'd have plenty of advance warning of any ozone problems, but now we've got a hole in our atmosphere that you could see from Mars," he told *The New York Times*. A 50 percent loss of ozone, he added, "is harder to label as just a computer hypothesis."

☐ ☐ ☐

Officials at the EPA were eager to address the ozone issue.

In January 1986, the agency released its new proposal for ozone protection called the Stratospheric Ozone Protection Plan. Under the plan, the EPA would increase research and provide information for the government to use in deciding whether to further regulate CFCs. Since the Vienna Convention had "failed to agree on any appropriate global control measures," the EPA would schedule a new round of international workshops on the issue. This ambitious workshop schedule would begin in May 1986 and would be sponsored by the United Nations Environmental Programme while the EPA began its own studies on the status of the CFC-ozone depletion theory. By November 1987, the agency announced, it would make a final decision on Phase Two CFC regulations. The EPA announcement made no mention of the NRDC's lawsuit or the Antarctic ozone hole.

The EPA's proposal alarmed CFC industry officials. They had succeeded in burying the issue for eight years, but industry leaders had new reason to dread their next battle over the chemicals: Sitting in the environmentalists' corner was a new EPA boss that seemed committed to resolving the issue—even if that meant sacrificing industry.

Lee Thomas, a mild-mannered person who spoke with a southern drawl, had been appointed EPA administrator in January 1985 after Ruckelshaus had resigned. Thomas seemed to view the ozone depletion issue as one that he could act on without other agencies opposing him. The time also seemed right to pressure industry. Although they would certainly balk at regulations, there was talk of renewing research on CFC substitutes. Businesses that used CFCs, air-conditioning manufacturers for example, were beginning to look at how the chemicals could be recycled to prevent emissions into the atmosphere. And CFC manufacturers were considering the traditional alternatives—such as sulfur dioxide, ammonia, and methyl chloride—once again, although they warned that those chemicals raised safety issues. New CFC substitutes, however, were gaining attention. Chemicals known as CFC-123 and CFC-134a were being touted by environmentalists as promising substitutes for air conditioning and refrigeration. Unlike CFCs 11 and 12, the new substitutes would break down in the troposphere, before reaching the ozone layer.

Thomas was briefed on the CFC-ozone issue early that spring during sessions to hammer out the agency's budget. The NRDC lawsuit also prompted his attention, and Thomas wasted little time in devising a strategy for dealing with the ozone issue both domestically and internationally. During agency briefings, Thomas reached three conclusions: (1) ozone depletion was an environmental issue that demanded attention, (2) the problem had a solution, and (3) the solution had to be implemented in a global fashion.

The hint of a fresh fight on CFCs drew a large, quarrelsome crowd at an EPA workshop on the issue in March. Thomas announced that the agency's plan marked a renewed commitment to the issue, but he was careful to add that regulations were not guaranteed. He cited the January NASA report as evidence that scientists did not understand what was happening to the earth's atmosphere. The ozone hole was proof that computer models were missing a few things, to say the least.

But for Thomas, the uncertainty factor in the CFC-ozone argument was not a major problem. Almost every environmental issue had some degree of scientific uncertainty. The most important thing to consider, he reasoned, was the consequences of inaction. And Thomas also felt quite certain about the cause of ozone depletion.

Thomas told the workshop participants:

> In the face of all this scientific uncertainty, one might ask why has EPA embarked on programs to assess the risks and to decide whether additional CFC regulations are necessary? Why not simply adopt a wait-and-see attitude and hold off a decision until depletion is actually confirmed? Let me address this question squarely. EPA does not accept as a precondition for decision, empirical verification that ozone depletion is occurring. Several aspects of the situation suggest we may need to act in the near term to avoid letting today's "risk" become tomorrow's "crisis."

Protecting the ozone layer, Thomas told his rapt audience, was an issue "as potentially vast as any I have to deal with as administrator of the EPA."

Thomas' words sparked angry reaction from industry representatives at the meeting. At a luncheon the following day, Richard Barnett, the chairman of the powerful industry lobbying group, the Alliance for Responsible CFC Policy, told workshop participants that he feared a decision about regulation had already been made.

"To say the least, we are troubled by the current strategy of EPA to hold a series of international and domestic conferences intended to build a consensus around the nature and severity of the (CFC) problem and major options for remedy," Barnett said. "We should remain focused on the stratospheric ozone protection problem."

CFCs were being unfairly singled out when, in fact, many trace gases combined with CFCs contributed to the overall picture of the stratosphere, Barnett told the *New Yorker.* Even the Antarctic hole could not be used as proof that CFCs were harming the atmosphere, he said. "Although the observed reductions in the ozone over the Antarctic region are real, the ozone levels return to near normal soon after the October springtime begins, and no plausible mechanism has been proposed to explain this phenomenon," Barnett said.

(As for Rowland's suggestion that heterogeneous reactions could be responsible for the Antarctic depletion, industry officials reacted in nearly the same manner as in 1974 when Rowland first postulated the CFC-ozone theory. "What a wild idea," industry scientists responded to him. "Have you published any of this?")

When it came to industry's argument over the scientific issue, Kevin Fay, a bright young man hired as a lobbyist for the Alliance, made its case in a convincing manner.

"There is an enormous amount of scientific uncertainty involved in this issue," Fay told reporters. "We definitely support continued research. But the scientific consensus we hear is that there is no imminent danger, that there is time to reach an international consensus. We do not concede that there is an environmental problem in terms of the imminent threat to civilization or human health."

In particular, Fay urged that the matter be handled in an international setting, saying that further domestic regulations would only cripple U.S. industry and would do little to affect the environment since almost three-quarters of all CFC emissions were produced on foreign soil.

"People say, 'The U.S. has to be a leader.' Well, no one followed us on the aerosol ban," Fay charged. "Only about three countries followed us. If we set new standards it may provide other countries with an incentive to sit back and say, 'Well, the U.S. is handling the problem.' It's truly a global problem. What we're saying is let's not commit suicide here in the U.S. from an economic standpoint."

Several industry scientists attending the EPA workshop presented research papers that highlighted the profound importance of CFCs to the world economy and technology, including one paper that cited the fast freezing of French-style green beans as one of the major contributions of CFCs to the quality of life. To many observers, industry representatives seemed to be growing desperate, trying to shout down the whisper of further CFC regulations by parading the virtues of the chemicals. But if CFC manufacturers were threatened by the EPA's new interest in regulations, they weren't about to concede defeat. Du Pont scientist and spokesman Donald Strobach pointedly informed chagrined environmentalists that the company had discontinued its research on CFC alternatives in 1980, so confident were they that the CFC-ozone depletion theory had died for good. The company was, in fact, building new production facilities in Japan and was planning a move into China.

According to Strobach, who was soon to leave his post defending Du Pont in the ozone war to return to laboratory work, the scientific evidence still didn't support the need for regulations. But some people at the meeting who were familiar with industry's usual response felt that Strobach, in private, was beginning to have his doubts about how long his company could maintain its present position. He could see that the issue was becoming more political. And the media was interested in the problem once again. For his part, Strobach did his best to discourage the sensational treatment of the issue. He insisted on speaking of the "ozone reductions in Antarctica" rather than the graphic "ozone hole."

□   □   □

June marked the start of ozone festival days in Washington. Several hearings and meetings were held to discuss ozone depletion making the subject "the talk of the town," according to one reporter.

For two days, on June 10 and 11, science butted heads with public policy in colorful fashion as reporters, scientists, government officials, and chemical industry leaders packed the chambers of a Senate subcommittee to listen to arguments on the depletion theory.

"This is not a matter of Chicken Little telling us the sky is falling," committee chairman John H. Chafee of Rhode Island, lectured hearing participants and spectators. "The scientific evidence is telling us we have a problem, a serious problem."

Chafee's intense interest in the matter was sincere. The Senator, who walked with a slight lurch but always seemed to be at the center of a whirlwind of activity, was consistently well informed on environmental issues. A Republican, Chafee shed his conservative views when it came to environmental matters. In his direct, abrupt manner, Chafee demanded that equal time be given to discussing what can be done about the problem. He urged the nation's lawmakers not to repeat mistakes that were made in the handling of the acid rain issue in the early 1980s when "research was a substitute for action."

The hearings made for particularly inviting newspaper coverage. Witnesses seemed to sense that the CFC-ozone battle was heading into an ultimate showdown and used their most forceful and, at times, emotional arguments.

The opening day of the hearings featured several accounts from scientists who supported recent warnings of impending disaster from the greenhouse effect. The rise in carbon dioxide and trace gases will have an earlier and more profound effect than first thought, several witnesses testified. The witnesses noted that a 1983 National Academy of Sciences report had stated that the greenhouse effect was a "cause for concern" but concluded that there was time to prepare for it. But less than three years later, signs of greenhouse warming had been documented.

"Global warming is inevitable—it is only a question of magnitude and time," NASA's Bob Watson told the committee.

By this time, the EPA had begun to assess the possible damage from global warming.

The greenhouse effect had become a policy issue in 1979 when four scientists reported to the government's Council on Environmental Quality that manmade gases would contribute to global warming. A small program on the greenhouse effect had survived Burford's tenure at the agency, and a few committed EPA staffers had succeeded in convincing some of the nation's top atmospheric scientists that the topic required immediate study. The issue became especially prominent in 1985 when Veerhabadrhan Ramanathan and Ralph Cicerone announced that other gases could surpass carbon dioxide—considered the primary greenhouse gas—in contributing to global warming. Based on that early warning, the EPA had gone on to predict a range of horrors that could occur should global warming become a reality.

According to the EPA, changes in climate would drive particular species to move or die if they were unable to cope with the rapid changes in climate. Crops in semiarid regions of the world would fail. The midwestern farm belt, for example, might die out as the best farming conditions evolved to the north in Canada. The number of days in which the daily temperature exceeded 100 degrees could increase from 1 to 12 in Washington and from 3 to 20 in Omaha.

The warming would also cause sea-level increases of 2 to 12 feet by the year 2100, the EPA warned, as glaciers melt, ocean patterns change, and water expands as the result of warming. A three-foot rise would easily flood such cities as Cairo and New Orleans. All coastal areas, on which almost half of the world's population lives, would be subject to flooding due to high tides and storms. Protecting the nation's east coast from a one-meter sea-level rise would cost an estimated $10 to $100 billion. Third World countries, of course, could not be expected to finance such protection, and areas such as the delta regions of South Asia might be ruined. Even in industrialized nations, protecting fragile coastal wetlands, drinking water, and beaches might be impossible.

While many uncertainties remain about the onset of a greenhouse crisis, NASA's James Hansen, a top authority on global warming, stunned the nation by reporting before the Chafee hearings and at various workshops that the earth had been warming about 1 degree in the past century. While 1 degree seems to be of minor importance, only a few degrees of cooling over a century can lead to an ice age. A few degrees of warming can lead to changes of equal magnitude. The increase was within the range of normal variability, Hansen said, but also "is of the magnitude expected as a result of increasing carbon dioxide and trace gases."

Although carbon dioxide had been feared as the major contributor to a greenhouse effect, Hansen reported that trace gases such as CFCs and nitrous oxide now contributed to the greenhouse effect at a rate comparable to carbon dioxide. Another contributor to global warming was the increasing destruction of the earth's rain forests. Called the "lungs of the earth" for their ability to remove carbon dioxide from the air, rain forests that are chopped down actually release carbon dioxide because of the rotting wood. Rain forests in many parts of the world are being destroyed for agricultural or industrial development. In other areas, the forests are being lost due to acid rain and demineralization of the soil.

But, Hansen said, "The greenhouse issue is not likely to receive the full attention it deserves until the global temperature rises above the level of natural climate variability. If our model is approximately correct, that time may be soon, within the next decade."

The greenhouse effect was just one consequence of the continuing use of CFCs, experts warned at the hearings and at a government workshop later that month. The earth's inhabitants would also pay for the conveniences of CFCs through increases in ultraviolet light.

Although much research remained, experts reported that too much ultraviolet light might damage DNA—the basic chromosomal material that transmits the hereditary pattern—of some species. Studies at the University of Maryland found that a 25 percent depletion in ozone could result in a 20 to 25 percent loss in total soybean crop yields. The UV radiation, it was reported, could reduce seed quality and lower the resistance of a species to pests and diseases. UV light might change a plant's rooting depth, allowing weeds to outcompete the plant for water and nutrients. While many crop plants had been screened for UV sensitivity, experts reported, little was known about the possible effects of increased radiation on native terrestrial plant communities which account for 90 percent of the world's plant productivity.

Experts also warned of damage to aquatic systems. According to R.C. Worrest of Oregon State University, experiments have demonstrated that radiation causes damage to fish larvae and other juveniles that are essential to the aquatic food chain. A 25 percent loss in ozone, scientists estimated, could result in a 35 percent loss of phytoplankton. Such an effect might have disastrous consequences on the world's fisheries. The Peruvian anchovy fishery, for example, is dependent on phytoplankton blooms.

Humans, too, would suffer greatly. Edward Emmett of the Johns Hopkins Medical School reported that ultraviolet light was suspected to cause DNA damage in humans, perhaps triggering changes in the human immune system. Diseases such as herpes and lupus might be triggered by increased exposure to ultraviolet light.

Even more was known about the effects of UV light on the skin and eyes. Experts reported that the sun's radiation was known to increase cases of skin cancer, including the dangerous melanoma, and contribute to the aging of the skin. Ultraviolet light is also responsible for damage to the eye, including the cornea, and contributes to cataracts and visual aging.

□ □ □

The consequences of losing the earth's ozone shield generated much interest now that an actual hole had been reported over Antarctica. Newspapers and magazines that summer graphically detailed the scientists' predictions and raised the question of whether it was too late to repair the damage.

Now an old hand at government testimony, Rowland used the June hearings to deliver an eloquent and emotional statement on the status of the ozone depletion theory. He declared that the statement he and Molina had made in their June 1974 paper can "with the benefit of 12 years of intensive study, now serve equally well as a brief summary of the facts of the chlorofluorocarbon-ozone problem.

"The Rowland-Molina paper in 1974 was an explicit early warning, and made the nature of the problem very clear. . . . Since that time, we haven't been looking for early warnings, but instead have been ignoring them and debating instead over how much ozone depletion we are willing to tolerate as a world society."

While the nations of the world could have adopted an attitude of prudent caution for the atmosphere, Rowland charged, "the governments of the world have instead adopted an attitude of prudent caution toward interfering with the chlorofluorocarbon industry."

Rowland pointed to the connection between the ozone hole and the rapid growth in CFCs in recent years. When heterogeneous reactions are added to computer models, he said, such deviations as the Antarctic ozone hole become possible. Due to the discovery of heterogeneous reactions, he said, "A plausible chemical explanation now exists not only for the enormity of the losses, but also for why they occur over Antarctica in the spring, and in the 1980s."

He challenged policy makers to consider whether they could look at the ghastly phenomenon and wait a few more years while questioning what to do about it. Rowland suggested a solution for them: "If our prime concerns are the atmosphere, the ozone layer, and the people it shields, the obvious answer is to discontinue this experiment without waiting for all of the answers."

Rowland's assertion that the Antarctic hole was caused by chlorofluorocarbons was particularly bold and risky. At least three major theories explaining the ozone hole (only one suggesting that CFCs were to blame) were being hotly debated, and no major experiments had yet been undertaken in Antarctica. Most scientists, even those who favored a particular theory, were unwilling to say they were certain of the cause. NASA's Watson chose to take a more neutral approach to the problem. While showing subcommittee members dramatic satellite film of the Antarctic depletion, Watson stated that its cause was a mystery.

"It is not yet evident whether the behavior of ozone above the Antarctic is an early warning of future changes in global ozone or whether it will always be confined to the Antarctic because of the special geophysical conditions that exist there," he told the committee.

Rowland was, once again, apparently alone and way out on a limb. But it was certainly a position to which he was accustomed.

Round two of the hearings, on June 11, gave Reagan administration officials their chance to defend the government's lack of action to protect the ozone layer. But there was one noteworthy exception in the administration's defense.

Receiving a final briefing on the status of the CFC-ozone theory the day before he was to testify, Lee Thomas went home that evening and made what several of his aides have called "a personal decision." The next day, Thomas tore up his prepared testimony and announced his belief that "some intervention" by government to address the buildup of manmade gases, specifically CFCs, now appeared to be necessary "even while there is scientific uncertainty."

Thomas was the only administration official to break ranks, and he encountered some forceful rebuttals.

William R. Graham, deputy director of NASA and the nominee for the new chief science adviser position, testified that perhaps 10 more years of study would be needed before enough of the scientific uncertainties were answered to require government action.

"Projections for the future have a large uncertainty to them and have to be reduced before we take actions for the future," Graham said.

Graham's lackadaisical attitude, later nicknamed the "go slow" policy, upset several subcommittee members who saw the matter as extremely important and in need of urgent resolve. Senator George J. Mitchell, a Democrat from Maine, eyed Graham as he left the witness chair and remarked sarcastically, "We all hope you're going to be a vigorous advocate for attacking this problem."

Mitchell later complained bitterly to *The New York Times*, "We are again faced with the dismaying prospect of an administration policy that everything must be known before anything can be done."

Representatives of the Energy, Commerce, and State departments backed Graham's "go slow" philosophy saying that hasty regulations on CFCs were likely to create major economic and foreign policy problems.

"Causes and effects of climate change are simply too poorly understood to warrant changes of energy policy," said Alvin W. Trivelpiece, director of the DOE's Office of Energy Research.

Chafee asked Trivelpiece why the United States should not be a leader and act unilaterally.

"The United States should take the position of leadership based on solid facts that are not disputed in the international arena," Trivelpiece answered.

"I have a little trouble following that rationale," Chafee retorted. "The United States is rich and can take the lead. If we wait for everyone, we won't get to first base."

Before the hearings concluded Chafee urged President Reagan to use his next meeting with Soviet leader Mikhail S. Gorbachev to discuss the greenhouse effect and a worldwide plan to prevent environmental catastrophe.

Reagan administration officials and the chemical industry agreed on one thing quite clearly: the need to draw other major CFC-producing nations into the bargain if regulations should become a necessity. Even Lee Thomas, asked by the *Washington Post* why the United States did not take a leadership role by considering additional CFC regulations on its own, replied, "They are not easy steps to take. There are social and economic disruptions."

The officials who had sat through the frustrating meetings leading up to the Vienna Convention saw few prospects to begin reducing CFCs. At the Chafee hearings, State Department official Richard E. Benedick noted that attempts to negotiate worldwide controls on CFCs at the Vienna Convention had ended in "total gridlock." The United States and the USSR had insisted on their own terms. Japan saw no need for intervention, and several Third World countries were merely not interested.

As deputy assistant secretary for oceans and international environment and scientific affairs, Benedick had been the U.S. representative at the Vienna Convention. A cool, deliberate man with a thin smile and a patrician air about him, Benedick had admitted during a speech at the March EPA meeting, "Looking back at the past round of negotiations, I believe that the international community may have put the cart before the horse: We are trying, in effect, to make a risk management decision before conducting a risk assessment." The next series of workshops would, he said, show why ozone protection was important.

Even if U.S. representatives attending the international negotiations were convinced of the need to act swiftly—which they were not—they had few allies. Ambassador John D. Negroponte, Assistant Secretary for Oceans and International Environment and Scientific Affairs, had told the Senate Foreign Relations Committee that spring that the Vienna Convention might serve to promote a deeper exchange of information among a group of suspicious and self-interested negotiators.

> For almost 10 years, we have not had reliable worldwide data on CFC production . . . Since 1976, the Soviet Union, a major CFC producer, has not supplied production data, despite requests to UNEP and the U.S. Chemical Manufacturers Association, which attempts to publish CFC data annually on a worldwide basis. However, the USSR participated in the negotiations and is a signatory to the convention. Other countries, such as China, may be producing CFCs, but they are not reporting this information. Once the ozone convention enters into force, perhaps within the next two years, it is expected that more countries, including the USSR, will provide CFC production data.

Negroponte urged the government to ratify the convention, but added, "It does not commit the United States to additional regulatory undertakings."

Although high hopes prevailed, an international workshop on CFCs held in Rome during the first week of May failed to differ from past meetings in one significant respect: Several nations sent delegates from the CFC industry. Participants also failed to agree on a range of estimates on future growth of CFCs and other trace gases which could then be used in computer models to give more accurate estimates of depletion. Such a task certainly wasn't an unreasonable starting point in discussing future CFC regulations. But, Benedick said, "I gather that the workshop's limited success was due in large part to the view of some delegations that because making future projections inherently entails uncertainties, nothing can be said about the future."

It was even difficult to gather the most minimal data from the delegates. UNEP representative Genady N. Golubev complained that he was running out of patience. At one workshop, Golubev said:

> Advocating patience is an invitation to be a spectator to our own destruction. . . . My concern here arises from a request by UNEP for factual data on current production capacity, production, emissions, use and regulatory measures concerning CFCs to all member states of the United Nations, all regional economic groups, and relevant non-governmental organizations. The purpose of the request was to make available necessary background information to the Workshop on Chlorofluorocarbons. The response was extremely poor. Only twenty relevant replies were received out of the more than 170 requests sent.

Golubev raised the question of whether the poor response could be attributed to ignorance over the huge and growing market for CFC products or simply to the low priority given the issue by states with more pressing environmental problems.

For several months, the EPA, environmentalists, and international negotiators had been discussing a freeze on CFC emissions, that is, stopping their growth but not asking countries to cut back on their use. But by June, a number of people were questioning whether a freeze on the chemicals would be enough. According to a report by scientist John Hoffman of the EPA's Stratospheric Ozone Program, CFC reductions of 85 percent were needed to stabilize chlorine in the atmosphere.

Hoffman's report stunned David Doniger at NRDC. Doniger, an attorney, had inherited the CFC problem from Miller and had given the lawsuit aimed at the EPA his full attention. A tall, thin man with sharp features and glasses, Doniger gave the impression of being committed, intense, and serious. He had a reputation as a workhorse and it seemed as if there were few environmental matters he couldn't discuss with knowledge.

Doniger, however, along with others, had assumed that if the emission rate were stabilized, the amount of damage to the ozone layer would also be stabilized. But Hoffman was saying that severe ozone destruction would continue under a freeze. Now the question was, is an 85 percent reduction even enough?

That question was at the center of discussions at the June 16–20 EPA-UNEP workshop in Arlington, Virginia, which drew more than 300 researchers and policy makers from 20 countries. Buoyed by the EPA's renewed commitment to the issue, the increasing interest in the greenhouse effect, and the recent publicity following the Chafee hearings, environmentalists used the workshop to kick off the strongest public campaign to resolve the CFC-ozone issue since the spray can wars. By this time, Doniger had concluded that a freeze was not enough. In his speech, cowritten with David A. Wirth, Doniger asked for a 10-year phase-out of the chemicals.

While Doniger and Wirth's plan struck many of the workshop audience as impossible, there was a distinct feeling that CFCs would have to be curtailed and eventually replaced if the issue were ever to be resolved. Michael Oppenheimer, an atmospheric scientist with an environmental group called the Environmental Defense Fund, demanded a program of research and action similar to the Manhattan Project that built the first atomic bomb.

"This is, after all, a mission to save the earth," he said.

Industry representatives were there to defend their interests, however. The Alliance's Richard Barnett, who was also the chief executive officer of a major air-conditioning company, warned that rushing into regulatory decisions could result in alternatives that threaten the safety of consumers and workers and have little or no environmental benefit.

"The Alliance for Responsible CFC Policy dismisses the notion of 'wait and see' on the ozone depletion and climate change issues as unacceptable," Barnett said. "I would hardly characterize the activities over the last twelve years as 'wait and see.' "

Practically everything that is known about the stratosphere has evolved over the past 15 years, Barnett said, suggesting that the Vienna Convention signaled "remarkable" progress. But, he said, "The science, as we currently understand it, tells us, however, that there is additional time in which to solidify international consensus."

While environmentalists and industry representatives stuck to their usual rhetoric, perhaps the most honest and startling assessment of the issue came from John C. Topping, Jr., the former staff director of the Office of Air and Radiation at the EPA. Topping, an engaging, direct person who later fulfilled his dream of running a private organization dedicated to solving environmental problems, told the gathering that the recognition of atmospheric problems had failed to generate any more resources to fight the problem.

"Despite indications that prospective climate change dwarfs all other environmental problems combined, fewer than nine EPA employees out of a workforce of about 12,000 work full-time on either the issue of greenhouse warming or stratospheric ozone depletion," Topping said. "Although the proportion has grown considerably in the past year, still only about one-tenth of one percent of EPA's total resources is devoted to addressing these issues."

Topping said that the problem was not due to the senior leadership of the EPA, "but to the statutory ambiguity of EPA's role and deep-seated hostility that the Office of Management and Budget has exhibited toward EPA's addressing such first-order environmental issues as climate change, indoor air, and radon. OMB has resisted giving EPA resources in apparent fear that the agency might uncover problems with resulting budgetary or regulatory consequences."

Although money was also given to NASA, NOAA, and the DOE for environmental matters, Topping said, there was no coherent federal focus on the problems.

The lack of a strong federal focus regarding an atmospheric protection plan was apparent. But in the spring of 1986, a few dedicated people representing the federal government's science arm were mustering their meager resources for what they considered to be one of the most perplexing and urgent scientific missions since the days of the Apollo space program. In what should have taken years to plan, a small group of scientists was being readied to make an unusual and risky expedition to Antarctica to live and work under the ozone hole.

10
Coldest Place on Earth
August 1986—October 1986

Susan Solomon received her wake-up call around midnight. Jolted into consciousness, she immediately began thinking about the journey that lay ahead. Elsewhere in the U.S. compound in Christ Church, New Zealand, 12 other American scientists were being roused out of bed after a very short night.

Gathering up her 35 pounds of survival and safety gear (which included an orange fur-trimmed parka and six pairs of gloves), Solomon joined the other scientists in a meeting room and sat down to await instructions from the pilot who would transport them this black, August morning to Antarctica—the coldest place on earth.

Solomon felt as if she'd already survived one small test as the unlikely leader of this expedition. The evening before, the group had been briefed about safety precautions for the trip. The navy pilot who addressed the scientists was enthusiastic, obviously enjoying his role in this high-profile mission.

"I understand you're the ozone expedition," he said, facing the group of 12 men and 1 woman. "I'm pleased to be taking you down to the ice. Who's in charge here?"

Solomon raised her hand and watched, amused, as the pilot's face registered disbelief, then surprise, then respect.

"Oh," the pilot stammered. "Good for you."

The group was scheduled to depart from Christ Church at 3:00 A.M. It would take nine hours to get to the American Antarctic outpost at McMurdo Sound. The pilots flying the Antarctic expeditions tried to land around noon when there would be a few hours of twilight. It was always safer, the pilot explained, to arrive in daylight should he need to make an emergency landing. Of course, the group might not even make it that far. Bad weather might force the pilot to turn the plane around—that is, if they hadn't yet reached the PSR. The PSR, or Point of Safe Return, was about four hours from New Zealand. Beyond that point, the pilot could not turn back because of the risk of running out of fuel. They would land no matter what the conditions.

Solomon had not heard of the ominous PSR before this night. And now, with the plane speeding down the runway, she began to think back on the events that had landed this congenial chemist on a somewhat dangerous two-month expedition to the world's most unfriendly environment.

The 13 American scientists and their sponsors had had only five months to organize this important expedition, and it seemed incredible to Solomon that they had actually pulled it off. It was in March, at an ozone conference in Boulder, Colorado, that Bob Watson of NASA, along with several scientists and representatives from the National Science Foundation, began to seriously consider an expedition to Antarctica to gather information on what might be causing the 50 percent loss of ozone in the region.

Although Joe Farman's paper was the hot topic of coffee-break conversation among the scientists, the Antarctic phenomenon was so new and peculiar, conference organizers in Boulder hadn't even scheduled any sessions to discuss it. Now Solomon, along with a few others, lobbied for the half-day session and one was hastily organized. But there was scant information on the atmosphere over the South Pole other than what Farman, Chubachi, and Heath had already discovered. The scientists were in desperate need of additional measurements if they were ever going to discover what was causing the depletion. Outside the formal sessions, rumors circulated about an expedition to the region. Watson, in his usual energetic style, worked the crowd of scientists asking, "Is there any chance we could do something this year?"

If there was one person who could bring the mission together, it was Watson. The director of NASA's stratospheric research programs, Watson was a master at blending the roles of bureaucrat and scientist. A fast-talking, energetic man who looked like something of a mad scientist with a thick, dark beard and wild hair, Watson was highly respected and knew his leadership was important in fostering cooperation in the atmospheric community.

There was some skepticism that an expedition could be organized so quickly. But Dave Hofmann, an atmospheric scientist at the University of Wyoming, was already committed to the idea. Hofmann was a veteran of Antarctica and loved the place. Since 1972, he had made trips to the region to measure such things as sulfuric acid droplets from volcanic eruptions. Normally, Hofmann flew to Antarctica in the late spring, around November, with hundreds of other U.S. scientists doing work there. This was the safest time to go, and during the months of November through February, McMurdo was a bustling, lively place, teeming with about a thousand biologists, geologists, chemists, and other scientists carrying out important research in the relatively pristine environment.

But the ozone hole presented a peculiar problem for scientists. Farman's data began on October 1 and showed the presence of the hole, while Heath's records showed the hole developed at the end of the Antarctic winter and beginning of the austral spring in September. Normally, the base was closed then. Only a skeleton crew of about a hundred people stayed on the base during the winter. And this rugged group remained with the knowledge that there would be no flights out, even emergency flights, until spring. The only flights made to McMurdo during the late winter were the so-called win-flys, or winter fly-ins. These were a series of flights made by McMurdo's gutsy support personnel to ready the base for the coming summer season. Using ski-equipped planes, win-fly crews prepared the icy runways for the scientists who would be following them in a matter of weeks. Win-flys, everyone knew, weren't generally used for scientific purposes. But if they were going to study the ozone depletion phenomenon, they would simply have to hitch a ride to Antarctica on a win-fly. Hofmann, who had made a few preliminary measurements of ozone during the summer of 1985, after Farman's paper had been published, was already planning to go early—even if he had to go alone.

Sending a larger group of scientists on the win-fly would be possible, Watson knew. A bigger question was whether the scientists—even the veteran Hofmann—could come up with instruments that would work well enough in the bitter conditions to make the expedition worthwhile this year. It was March. The instruments would have to be ready to ship in July. Looking around the group at Boulder, Watson knew that there were only a few who could pull the task off.

It was important for Hofmann to go, of course. He was the only person with experience in Antarctica. He also had instruments for measuring atmospheric compounds. Barney Farmer of NASA's Jet Propulsion Laboratory was another possibility. His group had a sophisticated balloon instrument that, Farmer told Watson, might be ideal for measuring particular chemical species in Antarctica. Phil Solomon and Bob de Zafra of the State University of New York had a ground-based device that could be used to measure chlorine monoxide at mid-latitudes. And, in the aeronomy lab at Boulder where Solomon worked, Art Schmeltekopf had been working for years on an instrument that could measure nitrogen dioxide. Perhaps, Watson suggested, the instrument could be used to measure chlorine dioxide. Detecting chlorine dioxide would be an indication of whether chlorine was playing a role in the ozone depletion. Schmeltekopf thought the instrument would work for such a purpose but couldn't make the trip in August. He suggested that someone else from his lab take the instrument to McMurdo. Solomon was as surprised as anyone else to hear herself volunteering for the mission.

"Oh, well, I'll go," she told Watson at the Boulder meeting. Later, she wondered if she had gone momentarily insane. Solomon created models on computers. She had never done any experimental work, let alone any field work in a place as inhospitable as Antarctica. And, she groaned, she would have to learn how to run Schmeltekopf's instrument. But she knew Schmeltekopf had created a solid and well-designed instrument that didn't require a skilled operator.

Feeling as though he was putting together a U.S. Olympic team for atmospheric science, Watson knew that these four groups and their instruments represented the best that they could offer at the time. Others concurred and suggested Watson persuade NASA to fund the expedition.

Returning to NASA headquarters in Washington, Watson and James Margitan, a young atmospheric researcher who was serving a temporary administrative post at the agency, began a round of phone calls to obtain permission to launch the expedition. The first task was to convince the people at the National Science Foundation, which organized all U.S. scientific expeditions to Antarctica, that the mission was urgent. Margitan put in a call to John Lynch of the NSF.

"We'd like to get some scientists to Antarctica to make some measurements," Margitan said.

"When?" Lynch said.

"August."

"Well, we want to do something in August of 1987."

"No, no, we want to do something *this* August."

"Impossible."

Within days, however, Lynch called back with a "well, maybe" answer for NASA and raised the possibility of the scientists going on a win-fly. NOZE I—National Ozone Expedition—would take place in less than five months.

□ □ □

Figuring out the logistics of the trip fell to Peter Wilkniss of the National Science Foundation. And, luckily for the scientists involved, Wilkniss was a veteran of Antarctic expeditions and was the right person to organize this eleventh-hour trip.

Wilkniss, too, had supported the idea of a 1986 expedition. As director of the NSF's Antarctic Program, he was responsible for overseeing the organization of American science activities in the region. The Farman paper had been a real bombshell. Wilkniss had heard someone mention the paper and had rushed out to buy a copy. A shirtsleeves type of administrator with a somber face and a husky voice, Wilkniss was appalled that all of the "multi-million-dollar fancy stuff we had up in the sky" hadn't seen the ozone hole. He was a scientist himself. In 1972, he had made measurements of CFCs in Antarctica with an instrument developed by James Lovelock. When Lovelock had heard Wilkniss was going to the South Pole, he lent the instrument to the American scientist, and Wilkniss had used it to measure a number of trace gases. The United States had maintained a strong scientific presence in the region for decades. And now Wilkniss felt his country had really goofed on missing the depletion. It was an awesome ozone loss, he thought. And there was no explanation for it. The findings had to be confirmed, he felt, and someone had better start trying to come up with an explanation. At the NSF, the ozone depletion expedition became the number-one priority for the upcoming Antarctic season.

Wilkniss and representatives from NASA, NOAA, and the Chemical Manufacturers Association (which supported the expedition and contributed some funding to it), met in Washington later that spring to plan the mission. Wilkniss supported the idea of sending the scientists in on a win-fly. The group discussed the list of scientists who should be asked to make the trip and decided to ask Susan Solomon to lead the expedition.

Returning to his office, Wilkniss picked up the phone and began making plans. The first step was to alert the logistics and flight people of the expedition. The scientists' equipment must arrive in Christ Church by August 1, Wilkniss instructed his staff. And the win-flys would be increased from six to eight to carry the scientists. The Americans' private support contractor in McMurdo, ITT Antarctic Services, was asked to assign a special coordinator to look out for the scientists' concerns and make sure they had beds, food, clothing, and keys to unlock the laboratories at the base. The expedition, Wilkniss said, was to be given the highest priority.

☐ ☐ ☐

Now, after months of intense planning, here they were—past the Point of Safe Return.

Solomon stared out the window as the plane neared McMurdo. The view was white, empty, and vast. Sheets of pack ice stretched into the horizon. If she looked hard, Solomon could sometimes make out snow-capped peaks and could see the wind whipping the snow. Everything was untouched and pure. She thought of the 1911 expeditions of those first Antarctic travelers, Scott and Amundsen. They didn't even have radios, Solomon thought with a shiver.

The reason they were making this treacherous journey was never far from her thoughts, however. If the ozone hole was being caused by a mere shift in circulation patterns, there was really little to worry about. A change in weather patterns would mean ozone was only being redistributed, not lost. But if chlorofluorocarbons were causing a permanent ozone loss, well, the very future of the planet and its inhabitants was at stake. Could this be a warning for the rest of the world? Although there were supporters of both theories, it was the latter that worried policy makers and generated intense public interest in the National Ozone Expedition.

The American scientists couldn't help feeling as if kleig lights were being focused on them even as they made their way through the polar night to McMurdo. The expedition had received considerable attention in the media and scientists who had never shown too much interest in the stratosphere were now suddenly devoting all their time to the Antarctic puzzle. Dozens of scientists would be waiting to see what the NOZE investigators found and whether the data implicated weather or chemicals or something else.

In September 1986, 13 U.S. scientists made a hazardous 2,100-mile trip from Christ Church, New Zealand, to the American base on Antarctica at McMurdo Bay to study why ozone levels were declining over Antarctica during the austral spring. Travel to McMurdo during this time of the year is considered hazardous.

"This is one of the most challenging things that we've ever come across in atmospheric chemistry," Solomon told *The New York Times* in a prominent article in July, adding, jokingly, that she would make the trip with her special, ultraviolet light-blocking sunglasses. But, deep inside, Solomon merely hoped her group would find *something*.

"On the one hand, it's very exhilarating and challenging," Cicerone told the *Times*, summing up the feelings in the science community. "And on the other, it's frustrating and scary—scary because it's hard to place your bets with any confidence."

The scientists were under a great deal of pressure to come back with answers, although people who were experienced in Antarctic science, such as Wilkniss, didn't really expect the group to generate much useful information on this first expedition. The scientists were charged with trying to detect what chemistry was occurring while the depletion was taking place. It is always difficult to measure the rates of particle reactions, and the conditions in Antarctica would make the task more difficult. Dave Hofmann's experiments might succeed, of course. He was familiar with the Antarctic conditions—at least its summertime conditions. But the others would be lucky if their equipment survived the trip and worked well enough to gather any data. The purpose of the expedition, the planners knew, was to introduce the group to the environment so that they would learn enough to return the next year with better skills and equipment.

"Nothing works the first time you go to Antarctica," Lynch, of the NSF, had told the expedition organizers. "So don't count on any of these things working."

Rowland, also, tried to lower expectations.

"When they get back, there is going to be demand for a yes or no answer (on whether CFCs were playing a role in the depletion)," Rowland said weeks before the expedition. "But the people who go to Antarctica don't usually get anything the first time they go."

Hofmann, perhaps, had more reason to be nervous than anyone. After all, he knew the potential problems the group could encounter just trying to arrive at the base in one piece. The most serious risk was that the plane could crash and lives would be lost. While the U.S. team wasn't greatly worried about such an event, Hofmann knew there was a good chance the scientists would have to land in a whiteout, a swirl of blowing snow that makes it impossible for the pilot to see the ground. During whiteouts, which Hofmann had experienced several times before, the pilot is forced to land the plane in a wide, open space, bringing his aircraft down slowly until its skis hit the ground. After arriving at the base, there was also the chance that the pilot would encounter great difficulties in returning to Christ Church, and the scientists had been warned that their instruments would have to withstand something called "drifting."

Drifting was required when the weather conditions were so bad that the pilot might have trouble taking off again. The back doors were opened as the plane landed, and as the aircraft jerked ahead, the cargo flew out the doors. It was simply the quickest way to unload the aircraft. But it was also a good way to ruin expensive equipment that couldn't necessarily be repaired in the limited environs of McMurdo.

Hofmann, however, wasn't particularly concerned. He loved Antarctica.

"In Antarctica, it's a challenge to do anything," Hofmann explained. "To get the balloon off is a challenge. To get the data analyzed with a limited amount of equipment is a challenge. And when you come out of there with a really big set of data, you feel like you've accomplished something."

Of course, on the return trip home, there were those days in New Zealand in which Hofmann could partake of his favorite beer.

NOZE I didn't start off particularly well, however. Almost halfway there, one of the plane's landing skis had frozen and the pilot chose to turn back. But they made it the next day, landing in 30-below weather and on runways that had taken nearly 15 hours to clear. Drifting wasn't necessary, but the plane was unloaded rapidly, and the equipment piled to the side of the runway so the pilot could take off again.

Staring at the ton of equipment sitting exposed in rapidly dropping temperatures, the scientists found they were not quite isolated. Support personnel from ITT Antarctic Services rushed out to greet them, excited at the arrival of the scientists who would be studying this horrible ozone hole that hovered over them. It was noon. By midnight, the last piece of equipment had been moved into a storage building. The equipment had survived the rough trip intact. Now it was up to the scientists.

Solomon arrived in her dormitory room after midnight, exhausted and cold. The room was sparsely furnished and, she noticed with dismay, didn't have particularly good shades on the window. Oh, well, she thought, removing her thermal underwear. Suddenly, she heard the unmistakable screeching sound of a foot pressing into the hard, icy snow outside her window. Solomon walked over to the dilapidated shade and peeked through a crack. There was a man outside.

A peeping Tom! she thought in amazement. At minus 40!

"It was the last place in the world I expected to find a peeping Tom," she told her coworkers the next day. "I shudder to think how long the guy was out there waiting. He must have been out there an hour. But, hey, it's a long winter."

Life in Antarctica was full of thrills and chills, the scientists were to discover. During their scant free time, when the weather was just too bitter to work, the group enjoyed watching a videotape of the Public Broadcasting Station's dramatization of Scott and Amundsen's expeditions called "The Last Place on Earth." Mealtimes were always

The snow-packed Ross Island, Antarctica, is shown from a high-altitude balloon used by scientists during the National Ozone Expedition in 1986. The picture shows the frigid desolation that has led to this continent's description as "the coldest place on earth." (Photograph courtesy of Dr. David Hofmann, University of Wyoming)

interesting. By the end of the mission, when the biologists were starting to arrive for the summer season, the NOZE scientists found themselves dining with people who were the world's leading experts on seals or top penguin authorities. The isolation—there were no phone calls, and letters from home took four weeks to arrive—sometimes made the days long. But the brilliant Antarctic sunsets more than made up for it.

The NOZE scientists were fortunate. Early into the two-month expedition it became clear that they could, indeed, gather some good data on this first expedition.

□ □ □

The first success came shortly after the group's arrival when Hofmann's crew from the University of Wyoming gathered outside the compound at 6:00 A.M. to launch the first balloon. The balloon was designed to

McMurdo Station, Antarctica, is shown from Observation Hill in October 1986 during the National Ozone Expedition. (Photograph courtesy of Dr. David Hofmann, University of Wyoming)

make measurements of ozone at various altitudes as the ozone hole passed over McMurdo. The instruments aboard the balloon would also measure ice and dust particles in order to test the dynamicists' theory. If the theory was right, the instruments would detect particles rising— the so-called upwelling. But Hofmann was concerned. He had launched hundreds of balloons in Wyoming but he had never sent a balloon into temperatures that are 90 below. He wondered if the fabric of the balloon—which was similar to the wispy stuff garments from the dry cleaners come wrapped in—would withstand the frigid temperatures. Strong winds would make the launch even more difficult. No one wanted to send the sphere, which was 100 feet in diameter, crashing into one of the buildings on the base. But, after some tense moments, the balloon cleared the compound and the telephone wires and disappeared into the mysterious clouds overhead. In all, Hofmann's group launched 33 balloons during NOZE I. The only glitches were the aching frostbite Hofmann and others suffered. Part of ballooning involves tying dozens of knots. And that, the scientists discovered, was impossible to do with Antarctic-strength gloves on.

High-altitude balloons such as the one shown here were used by Dr. David Hofmann's research group during the first trip to Antarctica in 1986 to study ozone depletion. The balloons were used to take air samples and measure various trace gases. (Photograph courtesy of Dr. David Hofmann, University of Wyoming)

☐ ☐ ☐

Susan Solomon's group from NOAA had it a bit easier. All they had to do was sit on the roof of a laboratory building, where there was no protection from the stinging wind, trying to direct moonbeams onto a mirror and aim the light through a hole in the roof. The light was then analyzed to detect various compounds. The group was looking specifically for chlorine dioxide (a marker for the chlorine chain reaction) and nitrogen dioxide (a radical that could inhibit the chlorine chain). Finding chlorine would support Solomon's theory that CFCs were playing a role in the depletion. Finding nitrogen would support the "odd-nitrogen" or solar theory that natural solar activity was to blame.

For their experiment to work well, the group needed a clear night and a full moon that was positioned low over the horizon. Those conditions occurred during a three-night period, September 16–18. And, with astonishing luck, the scientists later learned that the polar vortex—the swirling mass of winds that shaped and contained the ozone-poor air—rolled over McMurdo just two days before. The conditions were ripe for gathering good data.

At times, however, Solomon had doubts about the entire mission. The scientists hadn't had time to build a tracking system for their instruments, which was why they had sit on the roof holding a mirror to reflect the moonlight. It seemed like a ridiculously backward way to do science. Each member of the team, however, took his or her turn on the roof, and sometimes one of the kind ITT workers would volunteer to mind the moonbeams. Solomon usually operated the instrument inside the building, but one day offered to do the roof work. Thinking about the old-fashioned way in which they had to operate in Antarctica, Solomon held the mirror and squinted at the small pipe in the roof to make sure the light was being directed into the instrument. She'd been on the roof for quite awhile now and she was beginning to get cold. Tired of squinting, she tried to open her right eye and realized it was frozen shut. Using one eye, she clambered off the roof and went inside, later relieved to discover that she hadn't suffered any permanent damage.

In another lab, de Zafra and Phil Solomon, the researchers from the State University of New York, also had initial jitters. They had built a microwave instrument that, they hoped, would measure chlorine monoxide—evidence that CFCs are triggering the depletion. The detection of chlorine monoxide was crucial to the chemical theories. Chlorine monoxide normally reacts with nitrogen dioxide to form chlorine nitrate. Chlorine nitrate and hydrogen chloride were the reservoir species suspected of holding chlorine. According to the chemical theory, reactions on the ice particles of polar stratospheric clouds freed this chlorine from its reservoirs to attack ozone. In a matter of days, de Zafra and Solomon were confident that they were gathering good data.

Things were also going fairly well over at Barney's Barn. That was the name given to the hut that was hastily assembled by ITT Antarctica personnel to house Barney Farmer's $3 million instrument. Farmer's team, from NASA's Jet Propulsion Laboratory, used a device called the Atmospheric Trace Molecule Spectrometer to measure the composition of the upper atmosphere. A version of the instrument had flown aboard the space shuttle.

Farmer had volunteered his instrument for the mission, feeling that it would be ideal for ground-based measurements in Antarctica. But the JPL group faced two difficult problems in using the huge instrument on the ground. Farmer's first concern was that the expensive piece of equipment would be damaged during the trip. His second concern was that the cold temperatures and high winds would make it impossible to get the instrument to operate in a precise manner. The instrument was built to allow light to filter through a hole in a wooden box. The scientists then studied the spectra; that is, they analyzed the light for various chemical species. Farmer's group soon discovered that the huge temperature difference outside the box compared to inside caused severe optical problems in trying to track the sunlight. Technicians were able to solve that problem, however, and Farmer's group eventually recorded the finest spectra they had ever experienced using that instrument.

Surprised that their instruments were performing well enough to gather solid data, the scientists met weekly to analyze what they had found and try to fit the pieces into a computer. Farmer knew his group was getting excellent data but couldn't actually tell what they had until they reviewed the data on special computers at home. But, even without Farmer's data, by the end of the expedition the group had become convinced that chemicals were causing the rapid ozone depletion. There was some strong evidence against the other two major theories.

Initial evidence from Hofmann showed that aerosols in the vortex were sinking, not rising, as the dynamical theory proposed.

Susan Solomon's group found high levels of chlorine dioxide— a species that had never been detected before—which implicated CFCs. The NOAA group also found that nitrogen dioxide levels inside the hole were among the lowest in the world, confirming the New Zealanders' observations. This finding cast doubt on the odd-nitrogen theory, which predicted high levels of nitrogen dioxide and tended to support the chemical theory.

On October 20, the scientists gathered in a room at the base to relay this consensus, via satellite, to a gathering of reporters at the NSF headquarters in Washington. The press conference had been prearranged in order to satisfy the press and public's intense interest in the expedition. But, as everyone involved knew, the major question wasn't "What is causing the ozone hole?"; it was "Are CFCs causing the ozone hole?" Now, reporters waited for the verdict. The NOZE scientists still had much data to review and analyze upon their return to the United States. But the preliminary evidence pointed strongly to a chemical cause of the depletion, and Solomon, the spokesperson, felt scientifically and morally compelled to say so.

"We suspect a chemical process is fundamentally responsible for the formation of the hole," Solomon said, trying to be heard over static. The evidence, she said, was against the odd-nitrogen and dynamical theories.

Satisfied with the year's worth of work that they had squeezed into two months, NOZE I departed for the United States. But instead of receiving congratulations upon their arrival, the group encountered a storm of controversy.

11
Chlorine, Dynamics, and Snowballs from Space
October 1986—May 1987

Confused by the message they had received from Antarctica, members of the media had dutifully trooped off to ask the dynamicists what they thought of their theory being wrong.

The scientists who had proposed the dynamical and odd-nitrogen theories were aghast. They thought the evidence from Antarctica was far too preliminary to conclude that the ozone depletion was caused by chemicals. Solomon's statements during the press conference had, as one observer noted, "waved the red flag under the bull's nose."

Callis, whose odd-nitrogen theory had received the most damaging blow from the expedition findings, could not accept that this one, small expedition had nailed down the cause of the hole.

"Their suggestion that the solar cycle is not playing a role in this is wrong," Callis told *Science News* in October. "And even if it is not wrong, it's certainly premature."

Among the dynamicists, Goddard scientist Mark Schoeberl was particularly angry over the NOZE press conference. Perhaps because both were smart, eager, and outspoken, the debate grew especially heated between Schoeberl and Susan Solomon.

Schoeberl had been drawn into the Antarctic problem innocently enough. At Goddard, Richard Stolarski (who went to work for NASA after leaving Michigan, where he had worked with Cicerone) was fascinated by the Antarctic phenomenon and had plastered computer printouts of the data on the walls outside his office. Stolarski wanted to see an actual picture of the depletion, however, and asked Schoeberl to take the color satellite images of the hole and piece them into a computerized movie to show how the hole appeared each September then disappeared by late October. The movie, which dramatized the hole in a way that the public could clearly understand, was shown around the world and helped to focus attention on the seriousness of the phenomenon. While making the movie, Schoeberl began thinking about the data and began working with Stolarski on a dynamical theory.

Schoeberl had made a small expedition of his own in 1986. He had dug up temperature data from the National Meteorological Center near Washington which showed that while ozone levels had been dropping, temperatures in the polar vortex had declined, too. The finding that temperatures in the vortex had fallen as much as 18°C in the past seven years supported the theory that air movements were causing the depletion, Schoeberl said. The colder the temperatures, the greater the upwelling of air. It irked Schoeberl and other dynamicists that the NOZE scientists dismissed their theories as incorrect without addressing this evidence. As dynamicist Mahlman told *Chemical and Engineering News*, "Irrespective of whatever chemical changes are going on, there is considerable evidence that sometime after 1979 there was a change in the meteorological wave activity in the Southern Hemisphere."

If the NOZE scientists had chosen the Antarctic press conference to make their case, the dynamicists were going to have the opportunity to argue their side as well. That opportunity came about with a special issue of the November 1986 *Geophysical Research Letters*.

Early in 1986, the editors of the research publication had asked Schoeberl and Arlin Krueger of Goddard to edit the special edition. Schoeberl and Krueger agreed and issued a request for papers. They began receiving the first submissions in June. Schoeberl and Kruger were startled by the huge response, and in particular, by the number of papers that suggested the ozone-temperature correlation. After reviewing the papers, Schoeberl concluded that the upwelling theory might be right—that the ozone hole was caused by some sort of natural fluctuation related to cold temperatures. The papers generally supported the idea of upwelling, although some dynamicists wondered if the colder temperatures were a consequence of the ozone depletion, not the cause. One paper appearing in the issue, by Keith Shine of Oxford University, suggested this possibility. Although the paper was largely ignored, it was later acknowledged to be correct.

Schoeberl was somewhat put off that several chemists prominent in the Antarctic debate had declined to submit papers to the GRL. One chemist who had been a co-author of a chemical explanation paper that had appeared in *Nature* earlier that year had responded to Schoeberl's request by informing him, "We have nothing more to say." But the dynamicists had not had their say. And when Schoeberl sat down to write the overview for the special GRL issue he emphasized that upwelling might be the cause of the hole.

"Despite the number of public announcements, no clear link between manmade pollutants and ozone depletion over Antarctica has been established; indeed, a number of papers in this issue present serious alternatives to and constraints on the suggested chemical scenarios," Schoeberl wrote.

When Solomon had spoken at the Antarctic press conference, she had been unaware of other dynamical explanations that had been proposed during the two months she had been out of the country. She had simply tried to summarize how things stood based on what they had found in Antarctica and on the theories that had been submitted prior to the expedition. Still, the dynamicists felt she had chosen to ignore them. And now, with this thick and rather dynamical issue of GRL, the chemists felt the tables had been turned. The dynamicists were ignoring them and the expedition findings. With each side having made a surprise attack, war was declared.

In the next few months, the NOZE scientists and their supporters were accused of going to Antarctica with a prejudice that the chemical theory was right. With accusations and barbs flying (via press reports and over private telephone conversations), the argument snowballed throughout the remainder of that year and well into 1987.

The dynamicists continued to argue that NOZE didn't rule out their theories. McMurdo, after all, is situated at the edge of the hole where there is very little vertical air movement, making it difficult to disprove dynamical theories, they pointed out. NOZE was designed to test chemical theories, the dynamicists claimed, while the testing of dynamical theories required measurements taken from several sites.

Another problem was the chemical theories' reliance on polar stratospheric clouds to explain the rapid depletion through heterogeneous reactions. Satellite data showed that ozone was being depleted in an entire region to at least 45° south latitude—the southern tips of South America and Australia. But polar stratospheric clouds don't form in higher latitudes and, thus, chemical theories could not explain the losses there. Rowland, however, disagreed with this argument. He had long been concerned that heterogeneous reactions in higher latitudes could take place on other surfaces, such as dust particles. The mixing of the ozone-poor air from Antarctica with air masses throughout the Southern Hemisphere also could explain how ozone could be "diluted" at higher latitudes.

Chemical theorists, in turn, pointed to Hofmann's findings that aerosols in the vortex were sinking, not rising, indicating that air masses were moving down.

"The dynamical theory holds that ozone is supposed to be transported in rising air. But we don't see aerosols being moved up. The hole could be caused only by uplift if ozone were lifted along through 'immaculate transport'—transport that only moves ozone," Hofmann joked to *Chemical and Engineering News*. (In all seriousness, Hofmann found the debate between the chemists and dynamicists so disturbing that he eventually quit going to scientific conferences where the issue would be discussed.)

Back in Boulder, Solomon returned to work on the NOZE data wondering how the silly Antarctic press conference had turned into such a controversy. She blamed the media, in part, for blowing it up. Of course, some scientists had escalated the argument with some off-hand remarks. And it also bothered Solomon that people had made such a big deal over what was intended to be a summary of preliminary findings. Instead of hedging, the scientists were simply trying to be honest with their appraisals. But it was like doing science in a fish bowl.

"It takes awhile to check everything," Solomon explained. "You've got to allow some time to assimilate your results and that's something that a lot of people don't recognize. They think the answer just pops out of the computer."

In fact, some of the most important findings from the expedition weren't revealed until several months after the scientists' return.

For Solomon, the challenge was to find a way to analyze the data from Antarctica more specifically. A lot of the methodology for analyzing these kinds of measurements had been developed by a man named John Noxon. A few years earlier, Noxon had created a technique of doing sky-viewing measurements while working in the NOAA aeronomy lab and had found that nitrogen dioxide dropped off over the Arctic. He had committed suicide in January 1985. But as the days passed after Solomon's return from Antarctica, she found herself thinking about him, trying to recall snippets of conversations she had had with him about analyzing such data.

To analyze her data on chlorine dioxide, Solomon needed a spectrum of light outside the atmosphere to compare it to. But scientists didn't have a true measure of the compound outside the earth. One day in December, Solomon suddenly remembered something Noxon had said. He had told Solomon that one of the ways he found useful for looking at a lot of compounds was to compare spectra taken at twilight and noon, that this was a much more sensitive way of analyzing data. Solomon had expected to see chlorine dioxide in data taken in moonlight. But now, using Noxon's technique, her analysis showed evidence of the compound in daylight.

Solomon was thrilled by this finding. Detecting chlorine dioxide was important to chlorine theory. It was indirect evidence that heterogeneous reactions were involved because large amounts of chlorine dioxide could only be explained if heterogeneous reactions were breaking up the reservoirs of hydrochloric acid and chlorine nitrate. Detecting the compound in the daylight was even more significant because it confirmed the nighttime findings and matched the behavior of the compound suggested by models.

Solomon announced her daytime and nighttime chlorine dioxide findings early in 1987 but knew there was skepticism about the measurements. She looked forward to a March 1987 meeting at Boulder to discuss the NOZE data, now six months after the expedition. At the meeting, Solomon was surprised to learn that the State University of New York group, led by de Zafra and Phil Solomon, had also recently improved its analysis and had detected greatly elevated levels of chlorine monoxide. That discovery, along with Solomon's findings, the low nitrogen dioxide measurements, and the lack of evidence showing uplifting seemed to solidify the chemical theory explanation.

Schoeberl felt that the chlorine monoxide measurement was the strongest piece of evidence for the chemical theories. But there were lingering doubts. For one, measuring the compound from the ground was very difficult. Second, de Zafra's group had earlier reported a measurement of nitrous oxide from Antarctica that they later discovered was in error and had to retract from circulated preprints of their paper. Perhaps this chlorine monoxide measurement was wrong, too, and they would have to retract it, some dynamicists admitted thinking.

☐ ☐ ☐

By late that spring, everyone involved in the Great Debate had concluded that a more sophisticated 1987 expedition to Antarctica in August would be needed to draw any further conclusions. With better instruments and a year's worth of discussion behind them, the next expedition would give everyone another shot.

A number of different chemical theories were still in the running. Molina was pushing the chlorine dimer explanation, and Solomon was beginning to think he was correct. McElroy was still talking about bromine but was put off by the little respect the theory was receiving. Bromine was being ignored, he told *Discover* magazine, "because the political connection to CFCs got all the attention."

The odd-nitrogen theory, however, was dead. There was simply no evidence that the Antarctic depletion had occurred before—as it would have if the theory were correct. And none of the NOZE measurements showed high levels of nitrogen.

The dynamical theory, Mahlman told *Chemical and Engineering News*, "is still hanging in there. The quantitative aspects of the ozone hole can't be explained by either the chemical or dynamical theories. Both groups can come up with hand-waving explanations but neither can quantify them."

The chemical theorists were feeling pressure. If it turned out that the hole was caused by some sort of natural variation, the political movement to reduce CFCs might falter. A natural explanation in Antarctica wouldn't mean the Rowland-Molina theory was incorrect, but it could make people forget about that fact for awhile. A 10 percent ozone loss in 50 years just didn't have the same impact as a 50 percent loss in 10 years. While it would be good news for the earth if CFCs were not causing the ozone hole, NASA's Watson, who was also deeply involved in international negotiations on CFC regulations, worried that negotiations would fall apart if the chemical theories proved wrong.

Other people were also worried over "premature" statements that the hole was caused by CFCs. Even Ralph Cicerone felt compelled to keep silent.

"This was one of those isolated cases where I felt that we really did need two or three years to sort it out," Cicerone recalled later.

Cicerone was surprised at his own attitude. He remembered the many years that industry had asked for "a few more years" to do research—a phrase that had become a kind of buzzword for delaying CFC regulations. Now, here he was, saying that a few more years might be necessary. But Cicerone felt he must reserve his judgment in the year following NOZE I. Much more information would be available after the 1987 expedition.

In an interview in *Discover* magazine, Schoeberl vocalized the fears of many, warning the chemists of just how far out on a limb they were.

"The chemical industry is going to be saying things like, 'Well, you know, you guys were wrong. You guys really don't know what you're talking about.' And the next time we see a global change, they're going to say, 'Well, you guys were wrong about Antarctica.' "

The chemists had something else to be concerned about. If the dynamical theories were correct, it signaled that it would be much harder in the future to know how much ozone depletion was caused by CFCs. As it stood now, the 11-year solar cycle had to be factored in. What if air currents had to be factored in, too? Schoeberl warned of such a problem in his overview in the November *Geophysical Research Letters*.

"If a large part of the decrease in Antarctic total ozone is shown conclusively to be simply due to a change in the climate of the stratosphere, then it will become increasingly difficult to produce incontrovertible evidence of the chemical destruction of the ozone layer over the background of natural variability," he wrote.

But it was really pointless to speculate on such matters. Proof was what was needed. And both the dynamicists and chemists agreed that chlorine monoxide—ClO—was the "smoking gun" that would confirm the chemical theories. The next Antarctic mission must confirm or dismiss the presence of this compound.

"Every other place in the world where you have high ClO you have loss of ozone," Rowland told *Chemical and Engineering News*.

But not everyone had been heard from yet. In May, Goddard physicist Maurice Dubin and chemist Igor Eberstein presented a new theory at the American Geophysical Union meeting in Baltimore. They proposed that the ozone hole could be caused by cometlike snowballs from space. The scientists suggested that thousands of snowballs, ranging in size from inches to 30 feet in diameter, plunge to the earth's atmosphere each day, disintegrating and leading to the formation of the icy particles that destroy ozone. Despite the creativity of their theory, the two men had ignored other scientists' data and were soundly ridiculed.

"People always disagree with the competition," said Eberstein. "Like many new theories, it generates hostility from people who have established views."

That was one thing everyone who had ever worked on the ozone issue could agree on.

□ □ □

By the summer of 1987, Bob Watson had his hands full. The debate over the ozone hole was at peak intensity, and Donald Heath had come up with new information on global ozone depletion that was turning heads in Washington.

Earlier in the year, Heath and Watson had found themselves delivering embarrassingly conflicting testimony regarding ozone depletion to a congressional committee. While Watson stated that global ozone losses had not yet been detected, Heath testified, on the basis of his latest satellite data, that ozone fell about 4 percent over a seven-year period. The loss was far more than models predicted and many scientists objected to the data, saying that degradation of the satellite instrument had falsified measurements.

But Heath had some support. James Angell of the NOAA Air Resources Laboratory announced he had detected a 4 percent global decrease between 1980 and 1985 based on a new analysis of the Dobson stations. Angell attributed one-quarter of the loss to natural solar activity. But he told *Science* magazine, "I don't think there's any doubt there has been a decrease of a few percent in the last five or six years. This is the first time I've ever considered that we're seeing something. What it means I don't know."

Watson was uncertain what to make of the claims of global ozone loss. But one thing was obvious: The Antarctic ozone hole had alarmed people enough that any indication of global losses was bound to create more concern. The models had failed to predict the Antarctica depletion, critics charged. What if they were wrong about global predictions, too? Maybe scientists were underestimating ozone losses everywhere in the world. In any case, almost every measurement of ozone made so far that year seemed to be pointing downward.

After persuading his superiors at NASA of its importance, Watson organized a major study to review global ozone measurements. Called the Ozone Trends Panel, the report would feature input from 150 scientists around the world. By the end of the year, they would know if ozone depletion was becoming apparent, as Rowland and Molina warned would happen 13 years ago.

Watson underscored the importance of the study to *Discover* magazine. If Heath's data proved correct, he said, "it would mean our understanding of chlorine chemistry was totally, totally wrong."

In what was later credited to his supreme vision, Watson invited scientists from around the world to participate on the Trends Panel report. He felt that the involvement of several countries might mean that the results of the report—whatever they turned out to be—would be accepted by all countries. This, in turn, could assist in the international negotiations on protecting the ozone layer.

□ □ □

The first NOZE had yielded a surprising amount of good data, but Watson knew that many critical questions remained. A bigger, better expedition was needed this year and he had had incredibly little time to organize it. Still, he had promised that he would "throw everything" into the 1987 expedition. It would have ground-based instruments, balloons, satellites, and high-altitude aircraft. Planning with Margitan during the summer of 1986, the pair envisioned the 1987 expedition to be a miniature version of the famed Manhattan Project: the world's best scientists drawn together in a concentrated effort to find a solution to a life-threatening problem.

Watson was confident that the second expedition would resolve many of the uncertainties that lingered from NOZE I. He predicted that the scientists had a 90 percent chance of delivering a verdict by the end of 1987 on whether CFCs were to blame for the depletion. Others weren't so optimistic, saying it would take several years to reveal, conclusively, what was causing the Antarctic depletion.

Some people doubted whether such a large and complicated mission could be pulled off so fast. Even though he was new to NASA bureaucracy, Margitan knew that such projects normally took years to plan. If they pulled it off, the Airborne Antarctic Ozone Experiment would break all the records for how fast one could organize a government-sponsored project.

Luckily, funding didn't appear to be much of a problem, even though the expedition was going to cost about $10 million. The public's concern over the ozone hole, the countless magazine and newspaper stories on the phenomenon, and the two years of high-profile government testimony had more than convinced NASA officials that the project was worth the money. And, once again, the Chemical Manufacturers Association would be contributing to the venture.

A bigger problem was marshaling the people, resources, and equipment to carry out the experiments. The problem with NOZE I was that the measurements were all ground based and were taken at one site. Watson knew that better measurements of the chemicals comprising the ozone hole could only come from sampling the air at the altitude at which the ozone seemed to be disappearing—20 kilometers, or about 13 miles above the earth's surface. There was only one aircraft available that could fly that high, at the edge of the stratosphere, and that was the ER-2.

The ER-2 was Lockheed's version of the U-2 spy plane. Designed to fly at high altitudes, the plane was a delicate, jet-powered glider. Perfect for this mission, Watson thought. The aircraft had already been proposed for a January 1987 NASA project based in Australia to study the exchange of air between the stratosphere and troposphere. In the spring of 1986, Watson called his colleagues at NASA's Ames Research Center in Moffett Field, California, where the ER-2 was based.

"What's the possibility of taking something like the ER-2 down and making measurements over Antarctica?" Watson asked.

"You can't do it," an Ames official responded. "ER-2s don't fly over Antarctica."

It was no wonder that the Ames people were taken aback by Watson's request. The ER-2 was very sensitive to winds. It was difficult to take off with side winds of more than 15 knots or total winds of more than 25 knots. And the polar latitudes were notoriously cold and windy. With its one set of wheels situated under the fuselage, the ER-2 wasn't designed for icy, bumpy runways. It was also possible that the single-engine aircraft could freeze while flying in the −130° temperatures of the polar vortex. When it's too cold, Ames officials told Watson and Margitan, the controls of the ER-2 get stiff and sticky. In normal conditions, the pilot could simply descend into warmer air. But in the vortex, the air from 12 miles downward was all very cold, and the ER-2 wasn't designed to fly at lower altitudes. If the plane's engine were to stall at an altitude below 40,000 feet, the pilot couldn't restart it. And, finally, if the pilot were to ditch his plane over Antarctica, a search-and-rescue mission would be futile. The pilot was required to wear a pressurized suit while in the tiny cockpit, and there wasn't room for additional survival gear. Besides, even a wetsuit wouldn't protect a pilot from the frigid environment below.

Voicing those initial objections, the Ames officials acknowledged that the Antarctic mission was exceedingly important. They respected Bob Watson and knew he wouldn't make such a request unless he felt it was very important.

"Well," an Ames official told Watson, "it's not impossible, but it's pretty risky. It's really pushing things. You've got to convince us that it's important enough to do that—that this isn't a bunch of scientists off on a lark."

Like traveling salesmen, Watson and Margitan arranged for a meeting at Ames later that summer. Ames scientist Estelle Condon had already been asked to serve as project manager of the proposed airborne expedition, and Condon attended that first meeting along with a group of theoretical investigators. Watson, who chaired the meeting, asked the group the all-important question: If they were to go to Antarctica, what would they measure? If there was one thing Watson wanted to avoid, it was allowing the expedition to be governed by the available instruments. It was far better to decide what needed to be measured, then decide whether they had the instruments to do so. After a thorough discussion, the group concluded that they had the instruments to measure almost everything with the crucial exception of chlorine monoxide and bromine monoxide. But, Watson noted, Harvard scientist Jim Anderson had lots of experience measuring chlorine monoxide on balloons. Perhaps he could build an aircraft instrument. If Anderson could pull it off, the expedition would certainly yield enough data to make it worthwhile, the Ames group concluded.

The next tasks were to ask Anderson to participate, to apply for funding, and to ask the pilots of the ER-2 and DC-8 to undertake the mission. As the scientists left the meeting room, one looked at Condon and asked, "What do you think the reaction of the pilots is going to be?"

Condon was familiar with the aircraft operations people. She knew they were a hard-working and brave, but careful bunch. This was a much more hazardous mission then NASA typically asked the pilots to fly.

"I don't know," she answered. "My feeling is that they're not going to like this one bit."

By this time, however, Watson was heavily involved with the political part of ozone depletion; he was one of the United States' scientific advisers at the international meetings. And in his meeting with the ER-2 pilots and their managers, Watson made a convincing and patriotic case for the mission.

"Look, this isn't just a science problem," he told the group. "It's a problem that has important policy implications that a lot of people care about."

In a few weeks, after completing a detailed hazards analysis, the Ames officials and pilots consented to the mission. The ER-2 and a team of former Vietnam War fighter pilots would lend their services to the study of this great, lurking enemy.

The next problem was the delicate task of convincing some foreign country in the Southern Hemisphere to allow the United States to base the high-tech aircraft and a team of 150 scientists on its soil for almost two months. NASA had spent almost a year negotiating with the Australian government to fly the ER-2 out of that country for the stratosphere-troposphere study. In September, Watson began a round of talks with the embassies in Chile and Argentina to locate a base for the airborne mission. And, luckily, he didn't need to do a lot of convincing. While Americans were horrified by the ozone hole, few felt directly threatened since the phenomenon was half a world away and seemed confined there. But the inhabitants of the Southern Hemisphere were all too aware of the problem and wondered what might become of them if the hole grew, stretching out over their countries like fog rolling in from the ocean.

Finding the perfect place to base the expedition wasn't particularly easy, however. Condon, a cheerful, diplomatic woman with a good sense of humor, and several Ames officials obtained clearance that fall to visit several air fields in Argentina and Chile. The site had to fulfill several important criteria. Most importantly, the airfield had to have a hangar to house the delicate ER-2. It was also important that someone in the tower speak English. The town must be big enough to house 150 people.

Condon and her group inspected various airfields and found problems with many of them. Punta Arenas, Chile, however, looked promising. The civilian-military air field, called the Presidente Ibanez Airport, had a large hangar and good runways. The town had over 100,000 residents and had enough hotels and restaurants to support the U.S. contingent. The Chileans spoke English in the tower. There was just one big problem, Condon thought, looking over the air field. The taxiway from the hangar to the runway was atrocious.

Climbing aboard a military bus, the Ames group drove down the bumpy taxiway to get a better look. The taxiway was full of potholes, some of which were covered with huge sheets of metal. Finally, the group's chief maintenance officer got off the bus and walked slowly around the taxiway while Condon sat in the bus shivering. It was springtime in Chile, but she was colder than she'd ever been. Finally, the maintenance chief climbed on the bus and gave Condon a sad look. He stated somberly that the taxiway was so dilapidated he wouldn't even allow them to tow the ER-2 out to the runway. Letting the pilot taxi out was impossible. If NASA wanted to base the expedition here next August, he said, the Chileans would have to build a new taxiway.

Trying to build a new taxiway in less than a year seemed like a pipe dream. In the bus, Ron Williams, the chief pilot of the ER-2, said, "Estelle, we couldn't pave the *Ames* taxiway by next August."

"I know," Condon sighed.

But there was no other choice. Punta Arenas was the best place, and the mission was too important to postpone. Returning to her office at Ames, Condon sent money to the U.S. embassy in Chile to hire contractors who would excavate the old taxiway and pave a new one. She also arranged for the Chileans to build a parking lot next to the hangar and to have a laboratory and bathrooms built in the hangar. Then she crossed her fingers for luck.

Bob Watson felt things were falling into place nicely. There was just one more person to entice into the project. Watson knew this man would have the best chance at getting the crucial data that would reveal whether chlorine chemicals were causing the ozone loss. Watson found Jim Anderson in Idaho in August of 1986, enjoying a month-long vacation with his family. Anderson had been looking forward to the trip to relax and recover from his hectic schedule at Harvard. Watson's brief phone call left him feeling excited and yet burdened.

"This problem has gotten so serious," Watson told Anderson. "We've got to do something about it and rather quickly. I want you to build an aircraft-based instrument to measure chlorine monoxide and bromine."

They would need the instrument by next summer, Watson said.

"Well, the time is very short," Anderson said, informing Watson that he had never built an aircraft-based instrument before. "I'm certainly willing to do everything that I possibly can. But it is going to be a long shot."

"We've got to do everything we can," Watson answered.

Anderson understood Watson's predicament completely. The science community was in near hysteria over the ozone hole. People weren't responding rationally, Anderson felt. The only way to cut to the heart of the matter was to see if chlorine monoxide and bromine were present in the vortex and at what levels. If the chemicals were present, the depletion could become much worse. He told Watson he would begin work on the project after his return to Harvard.

It had been several years—since his revealing balloon flights to measure chlorine monoxide—that Anderson had been involved in the CFC-ozone issue. The past few years had been frustrating. Early in the decade, his research team had begun to develop a new generation of laser-based instruments to measure particles in the atmosphere. The instruments were much more sensitive and were designed to detect hydroxyl radical in the lower stratosphere. Hydroxyl radical was important in predicting ozone depletion. But, just as Anderson's group was finalizing work on the instruments, scientific ballooning in the United States took an unceremonious dive. In 1983, the factory that had been making material for the balloons changed hands and the old, reliable material became unavailable. Then, the management of the balloon facility in Palestine, Texas, was taken over by NASA and new managers, some of whom lacked long-time experience in ballooning. Problems ensued. From 1983 to 1986, Anderson was unable to launch a single balloon, and his work had almost come to a halt.

Finally, a new company began producing a reliable balloon material, and by 1986, the Harvard group was moving closer to the final stages of launching their laser system. Then Watson called. Anderson would have to put the balloons on hold again.

Anderson was given a budget of $400,000 and wasted no time in getting started. His first task was to call Art Schmeltekopf at NOAA. Schmeltekopf, who had designed the instrument Susan Solomon had used during NOZE I, was an expert on aeronautic experiments and high-altitude measurements. He agreed to brief Anderson's group on the art of building such instruments.

Next, Anderson called Lockheed to find out more about the ER-2 and its components. He also contacted a former college roommate, a fluid dynamicist at Ames, who was doing wind-tunnel studies, creating computer models analyzing flow patterns around aircraft instruments. Anderson convinced his old buddy to help him on a design. The instrument must be ultrasensitive—able to detect a tenth of a part per trillion of chlorine monoxide. But unlike a balloon-based instrument, an aircraft instrument had to deal with air flow. Free radicals of the kind Anderson was to measure could be destroyed while colliding with the instrument. If the instrument couldn't handle the air flow properly, he might not be able to get a clear, convincing result. He would end up having to apply corrections to the data which might cast doubt on its accuracy. And if there was anything Anderson feared, it was a wishy-washy result.

An undercurrent of anxiety ran through Anderson's team. After the ballooning fiasco the group felt it had let the scientific community down. For years they had been unable to contribute. This was their chance. Watson had shown enormous faith by asking them to build this instrument when, in fact, they had contributed little to stratospheric studies in the past several years. They had to make good on this request and deliver to the science community what they needed to know.

"There was no such thing as going in and coming back with some half-baked solution," Anderson said later. "In one sense, that is how science operates. You have to be very careful about interpreting information. On the other hand, a half-baked result would have been almost worse than no result at all."

By Christmas time, Anderson felt he had a design that could handle the fluid-flow problem. It was time to put his trust in the Lockheed engineers who would build the device so that it would work properly and not endanger the pilots. Blueprints were drawn up, and every single part—500 in all—had to be checked and fitted together for field trials by late May.

By early May, Anderson's team had compiled a huge box of parts that had never been assembled. They were scheduled to leave for Ames on May 20, where they would have two weeks to get the instrument ready and run it for the first time. Tension was mounting among Anderson's team. They knew that the crux of the airborne expedition depended on them. Everyone involved in the expedition recognized Anderson's role, and when Anderson arrived at Ames in May he was deluged with inquiries.

"How's the instrument working?" well-meaning colleagues asked him.

Instead of flattening their noses, which was his first reaction, Anderson collected himself and replied, "What do you mean 'How's the instrument working?' The instrument has never worked. It has just been converted from a basket of parts."

Finally, the test day arrived. The instrument was six feet long and occupied most of the area on the left wing of the aircraft. Because the ER-2 pilot would have his hands full just flying his craft over Antarctica, it was crucial that all the instruments were easy to operate. Anderson's instrument was computer controlled. All the pilot would have to do is hit a switch on the instrument panel and the computer would run the experiment. The first flight was intended to see if the instrument handled the tricky fluid-flow problem. After reviewing the data later that evening, Anderson knew that one major problem had been avoided. The next problem was to see whether they could detect chlorine monoxide. On the next test flight, they loaded the instrument with nitric oxide, which converts chlorine monoxide to chlorine. And, once again, the instrument passed the test. The data showed chlorine monoxide. Anderson was elated but exhausted. He knew nerves were frazzled. Many of his assistants were young, undergraduate students, and yet they had performed important roles in building the instrument, such as developing the computer software to run the experiment.

The next test would have to wait until the actual mission: knowing whether the aircraft could handle the load of instruments. Anderson worried about the risks of the actual flight. Still, the detection of chlorine monoxide had been a major hurdle and relief was almost palpable among members of the expedition research team. Anderson gave his team some time off. They would reconvene at Ames in early August to depart for Punta Arenas.

12
A Solution in Sight
June 1986—June 1987

Hopes for an international agreement to limit CFCs began to swell during the summer of 1986.

Despite the lack of a specific plan and complaints that "good will" was not enough, a spirit of cooperation was evident. The major CFC-producing countries had attended the UNEP workshop in Leesburg, Virginia, and the Soviet Union had released its production figures for the first time in ten years. The EPA's John Hoffman, who was charged with formulating the agency's proposals on ozone protection, was ebullient, telling *Science News*, "It was one of the most upbeat meetings (on ozone) I've ever been to."

It looked as though EPA's strategy was starting to pay off. The NRDC's Doniger, writing his assessment of the international talks later in *Issues in Science and Technology*, claimed workshops such as Leesburg paved the way for the negotiations by making progress in four areas:

1. Du Pont conceded that price, not chemistry, was the biggest deterrent to developing safe CFC substitutes. The fact that substitutes, such as CFC-123 and CFC-134a, were possible galvanized the U.S. political response.

2. CFC producers from Europe and Japan admitted that production rates were rising.

3. Negotiators agreed that chemicals built up in the atmosphere, thus facing the truth of the EPA's calculations that emissions

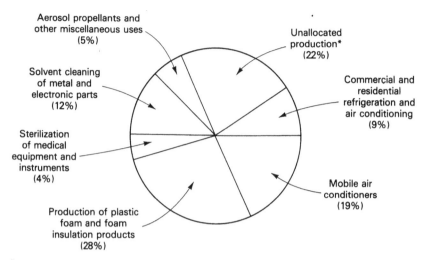

Aerosol propellants and
other miscellaneous uses
(5%)

Solvent cleaning
of metal and
electronic parts
(12%)

Sterilization
of medical
equipment and
instruments
(4%)

Production of plastic
foam and foam
insulation products
(28%)

Unallocated
production*
(22%)

Commercial and
residential
refrigeration and
air conditioning
(9%)

Mobile air
conditioners
(19%)

*Unallocated Production: This fraction represents the difference between total production and estimated usage in the categories shown. Factors in the difference include overseas trade, unreported military uses, and inexact accounting of sales. EPA is taking steps to obtain more accurate production and consumption data.

Various uses of CFCs have contributed to ozone depletion.

must be cut by 85 percent to keep the chlorine stockpile from growing.

4. It would be acknowledged eventually that the Antarctic ozone hole was caused by CFCs.

Hoffman, a young man with curly hair and glasses, seemed sincere in his desire to see the EPA take the right stand on CFC regulations. He was a capable scientist and downplayed his role as a bureaucrat. He felt the most important consensus was that the participants now agreed that CFC emissions were rising and that higher emissions required new actions.

"The whole argument that they weren't going to grow was completely wiped out," Hoffman later recalled. "Even more important than that was it came out that there were alternatives that could really succeed. That really just electrified the whole situation. There was a conversion process going on from believing that it was a bad environmental problem, but it would be a nightmare to solve, to it's a bad environmental problem, and maybe there is a good solution in sight."

Part of the EPA's plan was to try to implement domestic and international regulations in tandem. But mistrust had grown between the United States and other countries after the disagreement on aerosol bans. The United States' recent attempts to persuade other countries to adopt aerosol controls had been seen by some as a bullying tactic. Hoffman and others at the EPA viewed this as the wrong tactic. It was time to repair relationships with other countries on this issue and to enlist their help in coming up with a constructive solution to this problem that affected them all. After the hostile Rome meeting, Hoffman discussed the U.S. strategy with Thomas. Thomas, too, thought a constructive approach was best and authorized Hoffman to solicit a new, more amiable way to solve the problem.

The tactic worked. The Leesburg meeting went smoothly. Instead of trying to tell each other what to do, representatives from the various nations seemed willing to listen. Even the head of the European delegation remarked that it was time to stop arguing.

"At that meeting, I knew there was going to be a protocol," Hoffman said later. "It was only a question then of what the percentage reductions were and exactly what structure the protocol had."

□ □ □

Chemical manufacturers seemed to be overtaken by a new spirit, too. The dynamicists may have had a hard time believing that CFCs could be causing the Antarctic ozone hole. But, strangely enough, CFC manufacturers didn't even wait for the NOZE I scientists to return from their mission before announcing a slight change of heart.

On September 16, 1986, 12 years after Rowland and Molina's theory was published, the Alliance for Responsible CFC Policy held a press conference to announce it would support limits on the growth of CFCs.

The announcement was made by Alliance president Richard Barnett. Barnett had recently taken over control of the powerful industry group, and a few EPA officials and environmentalists viewed him as more open minded, more reasonable in his approach to the issue. But Barnett, in making his stunning statement, asked U.S. policy makers to be "reasonable," as well.

"Responsible policy dictates, given the scientific uncertainties, that the U.S. government work in cooperation with the world community to consider establishing a reasonable global limit on the future rate of growth of fully halogenated CFC production capacity," Barnett said.

Barnett cautioned that the policy shift did not mean industry officials believed "that the scientific information demonstrates any actual risk from current CFC use or emissions." But, he told *Chemical and Engineering News*, "The science is not sufficiently developed to tell us that there is no risk in the future."

Barnett cited NASA's recent estimates on depletion and the Antarctic mystery as reasons for the Alliance's policy change.

Kevin Fay, the group's public relations expert, emphasized that much scientific uncertainty remained. But, he told *Science News*, "We thought it necessary to develop a policy that provides some reasonable assurance that we don't ever get to the doomsday scenarios that have been the focus of so much attention lately."

Doniger had sensed that a change in industry policy was forthcoming. At the Leesburg workshop three months earlier, industry officials had indicated that some restraint on the rate of growth might be acceptable. And earlier in the year, a Du Pont official had given a paper at an EPA workshop stating that alternatives to CFCs were available but would cost more. If the price was right, however, manufacturers could do more research to develop new chemicals. Many workshop attendees viewed the paper as a sign from Du Pont that if a freeze on CFCs drove up the price of the chemicals, that would be a catalyst for supporting research on alternatives. Others were caught off guard by the announcement, however. Only a few months earlier, sales teams from Allied-Signal, the second largest CFC producer in the country, had been dispatched to encourage electronics manufacturers to switch to CFCs from other solvents.

Du Pont, however, was apparently looking over its shoulder at the ozone hole. A few days after the Alliance announcement, Du Pont announced that it would go beyond the industry position and called for a limit on worldwide CFC production based on the UNEP negotiations.

At the NRDC, environmentalists hailed Du Pont's action as "the biggest breakthrough since the EPA's 1978 aerosol ban." Du Pont officials based their policy change on

1. The increase in production of CFCs.

2. Models that showed that growth in CFCs of 3 to 5 percent per year would lead to significant ozone depletion.

3. The fact that CFCs contribute to global warming.

4. The inability of science to specify a safe long-term growth rate.

But the giant corporation's about-face bewildered many people and was viewed with some suspicion. The company finally seemed to be saying what environmentalists had been saying all along: that prudent action was required in the face of doubt.

Why was industry acting now? There were several theories.

There was the threat of regulation. The EPA was being forced— through the NRDC's lawsuit—to provide a plan for further restrictions. And the Vienna Convention, if nothing else, ensured that negotiations on worldwide CFC controls would continue.

Perhaps industry representatives were starting to believe the mounting scientific evidence that the chemicals could lead to rapid depletion. The ozone hole—even if its cause was still being debated— had rocked industry. The crisis had created enough public concern that chemical manufacturers might have felt their public images were being threatened. Du Pont, for one, had maintained a good safety record. It certainly didn't want to give the public the impression that it was callously sending chemicals into the atmosphere that could doom the planet. And it was possible that industry officials had received word that the NOZE I scientists were finding evidence that could indict chlorine chemicals.

There were also rumors that the CFC-ozone issue had been taken out of the Freon Division's hands at Du Pont and turned over to senior corporate management after the discovery of the Antarctic hole. At this level of the company, the issue would be looked at in a broader manner. CFCs were only a small part of this huge corporation's business, but trouble with CFCs could cast a shadow on the entire operation.

There were, perhaps, economic reasons for industry to act. There was increasing talk of ozone depletion leading to increases in skin cancer cases. If tobacco manufacturers could be sued for contributing to individual cases of lung cancer, couldn't CFC manufacturers be vulnerable to lawsuits over skin cancer cases if the chemicals were, eventually, found to cause depletion?

And there was the issue of CFC substitutes to think about. What if regulations—either international or domestic—were implemented? The first company to come on the market with CFC substitutes stood to profit handsomely. There were rumors that the Japanese had already begun research on substitutes, eager to find an alternative to the solvent CFC-113 in order to protect its billion-dollar electronics industry.

Realizing they might already be entering the game too late, the CFC industry began an earnest search for CFC substitutes in 1986.

Substitutes for aerosol sprays had been relatively easy to come up with in the mid-1970s. But industry officials knew that replacing the tons of CFCs used in refrigerators, air conditioners, solvents for electronic equipment, and in the making of polyurethane foam would be difficult and expensive. There were three main chemicals that would have to be replaced: CFC-11 was used in the making of polyurethane foam and in refrigeration; CFC-12 was used in refrigeration, air conditioning, and foam blowing; and CFC-113 was the marvel of the electronics industry for its use as a solvent.

There were several problems. CFC-22 was known to be a safe replacement for home air-conditioning units. The chemical contained a hydrogen atom that caused it to break up before reaching the stratosphere. But CFC-22 wouldn't work in automobile air conditioners or refrigerators because it required higher temperatures and pressures. Putting CFC-22 in cars would reduce mileage rates. One possible substitute for CFC-12 was CFC-134a, but it was expensive to make and it hadn't been tested for effectiveness or safety. There was some use of pentane as a substitute for foam blowing, but pentane was flammable.

With all these problems, industry officials claimed it would take at least five years to come up with any substitutes. But, said Barnett during government testimony, it was difficult to know just what progress had been made on substitutes. Manufacturers, in trying to position themselves for a competitive advantage, weren't too forthcoming about their research efforts.

Manufacturers also warned that replacements would hit Americans in the pocketbook. Worldwide CFC production was nearly 700,000 tons a year, and the chemicals sold for about a dollar a pound. Restrictions on the chemicals were sure to drive prices up. And research on substitutes required money. Du Pont announced that it had already spent $15 million on research. Allied officials said they had spent $10 million. These figures did not impress pro-regulation forces who pointed out that leading U.S. industries routinely spend a percentage of their earnings on research. But Du Pont official Joseph Steed told *Chemical and Engineering News:* "The development of alternatives is going to happen at a rate that corresponds to the amount of pressure that is applied. Right now there is no economic or regulatory incentive to look for other routes."

□   □   □

One of the first things Lee Thomas did after joining the EPA was to order a two-year study on the ozone issue. Thomas wanted a summary showing what would happen if CFCs were cut back by various degrees or if the agency simply did nothing. The report, released in the fall of 1986, showed the direction the EPA would have to take to comply with the Clean Air Act.

The analysis, the first comprehensive report on ozone since the NASA-UNEP assessment, showed that even halving the growth of CFCs 11 and 12 would not be enough to prevent serious damage to earth. Increases in ultraviolet light due to ozone depletion would produce 60 percent more skin cancers and 600,000 more cases of cataracts in the United States. Global warming, of which CFCs contributed up to 25 percent, would increase sea level and threaten crops. Doniger said the report shows "the ozone situation is more dangerous than we thought it was."

For months, Doniger had been trying to persuade the EPA to ask for a phase-out of CFCs instead of just holding emissions of the chemicals steady. At a meeting with EPA and State Department officials organized to hear opinions on the matter, Doniger again stated the NRDC's position that a freeze was not enough. But during a break in the meetings, Doniger was informed by a top-ranking NASA official not to get his hopes up. No one was seriously considering a phase-out, he was told. It was simply too bold a proposal.

Thomas, however, apparently agreed with the NRDC's logic. On November 5, the EPA and the State Department revealed their plan for worldwide CFC controls in a letter from the State Department sent to all U.S. embassies. The U.S. position included:

1. A near-term freeze on the emissions of all fully halogenated CFCs at or near current levels.

2. A long-term phase-out of the chemicals.

3. A periodic review of what is known about ozone depletion with revisions to follow if necessary.

Doniger was pleasantly surprised by the government's actions. But he was not yet ready to applaud the EPA's position. Instead, the NRDC pushed for a more specific plan: a 30 percent cut in CFCs in 18 months through a ban of aerosols and the use of CFCs in blown foams for food containers (considered a frivolous use of the chemicals) and a phase-out of the chemicals within ten years. Even if other countries refuse to act, the United States should implement domestic restrictions and prohibit the import of CFC-derived products from foreign producers, Doniger argued.

The NRDC plan received enthusiastic backing from environmental groups around the country. And although it was viewed as too radical by the EPA, the plan eventually became the basis for the official U.S. position as international negotiations resumed in late 1986.

In the tough world of international law, Doniger regarded the plan as a good bargaining strategy. Ask for more and maybe you'll get something that's a little less but is still acceptable. But, Doniger noted, once the U.S. policy was announced it became the official policy whether it was a platform for bargaining or not. "You can't turn around and say publicly, 'Hey, that wasn't our real position,' " Doniger later explained. That point became extremely important in the ensuing months when some administration officials began attacking the U.S. position as too extreme. The fact was, the official position had been announced.

On December 1, 1986, 17 months after the Vienna Convention, negotiations resumed in Geneva. The United States proposed restricting the chemicals to current rates and reducing emissions by 95 percent during the next decade—the crux of the NRDC's plan.

The European Economic Community and Japan protested. They would only consider a freeze on the chemicals, not a phase-out. Negotiations broke down when the United States refused to back down on its proposed timetable for reductions.

The Geneva meeting generated mixed feelings. The U.S. team felt that it had achieved a consensus that actions were needed to protect the ozone layer. One British official told *Newsweek:* "We're pretty well agreed about what to do now. We're still far apart on what to do later on." France's negotiator, Maurice Verhille, told the magazine: "There is no scientific evidence that suggests we should panic now. So we don't intend to panic."

To Doniger, who attended the Geneva meeting as an observer, it was apparent that Japan and the EEC countries were simply lagging in their understanding of environmental issues. Different economic conditions prevailed for CFC manufacturers in Europe, also. While there were a handful of major U.S. producers who could survive a phase-out of the chemicals, there were many smaller industrial uses of aerosols in Europe and those businesses couldn't survive a phase-out. He knew efforts to convince the EEC members of the importance of reductions would be difficult. But what Doniger and others didn't anticipate was that they would soon have to convince key members of their own government of the same thing.

☐ ☐ ☐

Word that the international ozone negotiations hadn't gotten off to a particularly smooth start alarmed members of the Senate Environmental Protection and Hazardous Wastes Subcommittee.

In hearings in the Dirksen Senate Office Building on January 28, Senator John H. Chafee criticized the home team for "backing off from its original position."

"There was obvious disagreement among delegation members," Chafee said.

U.S. Ambassador John Negroponte objected to Chafee's suggestion that the U.S. position had been watered down. But during further questioning from Senator Max Baucus, Negroponte admitted that a phase-out of up to 20 years might be required—twice as long as the U.S. had originally proposed.

"That is a major change from what I understood the U.S. position to be as late as December when privately we were being told that the period the delegation was authorized to negotiate was something on the order of 8 to 12, or maybe 14 years," Doniger responded in his testimony before the committee. "It is an odd policy to have your strong position as your secret one and your weak position as your public one."

Baucus was critical of several of the negotiation teams. He blasted the EEC for showing up in Geneva without a position and without any power. He charged the Soviet Union and Japan with lacking good will and pointed out that the U.S. government could hardly call itself dedicated to a successful international ozone protocol by slashing its financial contribution to UNEP. The U.S. contribution to UNEP had fallen from 36 percent of the organization's total funding in 1973 to 22.7 percent in 1987. The contribution was cut 6 percent in the last year, Baucus pointed out.

Chafee and Baucus were furious. Given the apparent inefficiency of the international negotiations, the senators said, they would proceed with plans to introduce a bill for a domestic phase-out of CFCs.

The senators' frustrations were shared by some of the scientists. Dan Albritton, a level-headed and highly respected atmospheric researcher at NOAA, had been appointed, along with Bob Watson, as a scientific adviser to the U.S. delegation. The Geneva negotiations had both worried and saddened him. It seemed as if the proceedings had everything to do with politics and little to do with science. A kind man who was exceedingly patient, Albritton wasn't exactly looking forward to the next round of talks and wondered if he should have politely declined this "honor."

Scientists were often aghast at the shortsightedness of the men and women who set environmental policies that the entire world would have to live with on a long-term basis. And the CFC-ozone issue was often used as a case in point. Science must be conducted free of political pressure, Wallace Broecker, a geochemist at the Lamont-Doherty Geological Observatory in Palisades, New York, had testified in an earlier Senate hearing on ozone depletion.

"The responsibility for basic research has been thrown to mission-oriented agencies which are pushed to do things immediately," Broecker said. When scientists apply to do long-term research, they are often told that the government needs short-term answers or lacks the funds for long-term research. NASA was a target of the respected scientist's remarks.

"NASA should be spending far more time looking down at the earth than looking at Uranus," he said.

The January hearings, while unpleasant, seemed to clear the air. At the next round of negotiations, in February in Vienna, the U.S. team proposed a 10- to 14-year phase-out. While their opponents were far from agreeing to a total phase-out, the EEC softened a bit and proposed 20 percent reductions of CFCs 11, 12, and 113. Japanese delegates, however, strongly objected to their precious CFC-113 being subjected to any reductions. And Japan, the EEC, and the Soviet Union all voted against reductions in the use of halons, chemicals used in fire extinguishers and, more importantly in their eyes, in some military applications. Halons, while used in much smaller quantities than CFCs, have a greater ozone-destroying potential, and the United States had insisted that the chemicals be discussed in negotiations. A Rand Corporation study showed that the use of halons was expected to double by the year 2000.

But if Lee Thomas' negotiating team was making progress, it failed to impress some members of the Reagan administration. Apparently, the position the EPA and State Department had announced the previous fall had slipped by some members of the administration unnoticed. But now, with some nudging by the chemical industry, officials in the departments of Interior, Energy, and Commerce began to balk.

The chemical industry had taken Chafee and Baucus' threat to push for unilateral controls seriously. "If the U.S. takes unilateral action, it takes the pressure off the rest of the world to act," Barnett protested to *Chemical and Engineering News.*

CFC manufacturers felt things were happening much too fast. When they had agreed to limited controls on CFCs last fall, they had done so with the idea that EPA wanted production of the chemicals frozen—something industry could grudgingly live with. But now EPA was pushing for something more—a rapid phase-out.

Conservative members of the Reagan administration also wondered if Thomas wasn't being overly aggressive. Very few countries supported the U.S. plan, some White House staffers pointed out to Thomas in private conversations. Why was the United States so much more concerned about CFCs and ozone damage than other countries? Others don't appear to be as worried, Thomas was told. What proof do you have that there is a problem? Why a phase-out instead of a freeze?

What went on during private, high-level administration talks that spring may never be known. But during a series of interagency meetings in April, Thomas—who was rumored to have wanted CFC reductions of 90 percent—may have been forced to settle with a position that was less than what he had hoped for. According to Doniger, who followed the proceedings as closely as he could manage, representatives from the Council of Economic Advisers and the U.S. Trade Representative supported the EPA and State Department while members of the Interior, Commerce, and Office of Management and Budget opposed the U.S. position. The latter faction simply doubted that the earth's ozone shield could be damaged so severely and so rapidly as to warrant the collapse of a major American industry.

Doniger and other environmentalists reminded the administration that it had already announced its official policy and to back down now would look weak. At one meeting, an administration official proposed an abrupt change of policy. But Doniger challenged that rationale.

"You're forgetting there already is a policy," Doniger stated. "It was announced in November. It was set forth on the table in December. It was reiterated in February at the negotiations. It has been reiterated in several sets of congressional hearings. And if you change this policy, you better make sure that everybody knows who did it and why."

The NRDC campaigned hard for the United States to retain its position. In a crafty move, the environmentalists called Vice President George Bush's office and reminded Bush's presidential campaign staffers that an international ozone policy would be about the only good thing the Vice-President could claim for the administration's environmental policy. Perhaps, with the right moves, the Reagan administration could become known as the administration that saved the ozone layer. This would be a fine feather for Bush's cap in his bid for the presidency. Bush's people acted interested in the NRDC's suggestion. But Doniger never found out if Bush's office exerted any pressure in pushing for an international agreement.

When the third round of negotiations resumed in Geneva that April, the United States offered a plan of 50 percent reductions. The chemicals would be frozen by 1990 with a 20 percent cutback in 1992. An additional 30 percent reduction would be implemented in the 1990s pending further discussion. But at the next round of talks, in May, negotiators constructed a number of possible agreements ranging from 20 to 50 percent reductions. Without claiming the position as its own, the United States' position for 50 percent reductions remained intact.

Gradually, other countries began to soften. For several months, environmental groups and scientists in Europe and on other continents had been lobbying their governments for CFC controls. While the European community was beginning to accept the idea of something more than a freeze on emissions, however, members of the U.S. delegation continued to receive pressure from the Reagan administration to settle for a freeze. But that was before the NRDC made its final tactical move.

Early that summer, Doniger heard that Interior Department officials were designing a program to convince the American people that a freeze on CFCs was quite enough. Ironically, the Reagan administration's "Personal Protection Plan" became the biggest joke of the summer and backfired just enough to give the U.S. delegation the chance for a strong ozone protocol.

13
Personal Protection
June 1987—September 1987

Doniger had grown angry over some Reagan administration officials'
attempts to derail the original U.S. position on CFC reductions. He
thought the reasons that the Interior Department had given for being
involved in the issue were silly. Interior Department officials had stated
that the country's position on ozone depletion was important to them
because halons were used on offshore oil rigs. On another occasion,
they claimed that because the Interior Department controlled more
land than any other government department, excess ultraviolet light
would affect their holdings more.

Doniger suspected that Interior Secretary Donald Hodel had not
wanted to go beyond a freeze in chemicals. But, one day in mid-
summer, Doniger was informed by a source that the Interior Department and the president's science adviser were making the argument that "personal protection and life-style changes" would suffice
instead of CFC regulations. It was an alternative policy that could be
presented to the president.

Seizing the opportunity, Doniger called several reporters and
briefed them on the rumored "Personal Protection Plan." The reporters dutifully called the State Department and other administration
officials and received confirmation that a "Personal Protection Plan"
was being advocated. Hodel had not specifically called for the use of
sunglasses, sunscreens, hats, and staying indoors and neither did Doniger. But, before long, someone had suggested that that must be what
the Interior Department meant by "personal protection" and "life-style
changes."

Doniger then pointed out that if each American had to buy two bottles of sunscreen per year, a hat, and a pair of glasses, the bill would come to $40 per person per year. That would result in an $8 billion national expenditure.

"It would be a lot less expensive to control the pollution," Doniger said. "Besides, you couldn't get the fish to wear sunglasses."

News of the "Personal Protection Plan" traveled fast. Within days, Hodel's remarks became a symbol of the Reagan administration's stubborn, dangerous, and weak position on ozone depletion. Cartoonists had a field day with the situation, and environmentalists sarcastically asked how Americans would go about putting sunscreens and visors on cows and stalks of corn since animals and crops would also be affected by an increase in ultraviolet light.

Hodel's comments succeeded in focusing new attention on a serious solution to a serious problem. Only a few weeks later, a resolution calling for CFC reductions of at least 50 percent, with a program for additional phase-outs attached, passed the Senate in a vote of 80 to 2. A cabinet meeting with the president followed during which the U.S. team was instructed to seek 50 percent reductions on CFCs and a freeze on halons.

There was even word that some cabinet members were now pushing for the 50 percent phase-out position. Evidently, some officials realized that a weak international agreement, which would also become the domestic position, might make it hard for the United States to defend a lawsuit. The more that was agreed to by other countries, however, the harder it would be to argue that the United States was not complying with the Clean Air Act. In order to protect U.S. chemical manufacturers, it would be best not to place them in a position of having to reduce the chemicals faster than other countries.

Whatever the reasons for the administration's eventual decision, Lee Thomas applauded the U.S. position that he would take to Montreal with the support of his president. Thomas said of the months of administration in-fighting:

There was quite a large discussion. . . . But as we went into the Montreal Protocol I thought we went in with as strong a position as we could have asked for. We went in with a full commitment that we wanted to cover all CFCs and halons, which was something I had really pushed for. We went in with a commitment that we wanted an early reduction process. We wanted not to just go for a freeze. We wanted a commitment on reductions and we wanted them actually to reflect a phase-out. We went in domestically with a very strong position. As a matter of fact, I, by far, had the strongest position going into Montreal than any other minister there, and I had it from the president of my country. And very few of the others had that strength.

☐ ☐ ☐

CFC manufacturers had won a temporary reprieve. It became apparent by May that the United States would probably not act on domestic CFC regulations without the support of an international protocol. U.S. Ambassador Richard Benedick was adamant about delaying U.S. actions until a protocol was signed. "We must be sure that our actions domestically support, and do not undercut, that international process, since this is clearly a matter which the U.S. cannot resolve alone," he told lawmakers.

Environmentalists were willing to give the U.S. negotiating team a little more time as well. The EPA was scheduled to issue a new regulatory proposal on CFCs by May 1 and final regulations by November. But on May 1, the agency, in cooperation with the NRDC, asked for an extension of that deadline to allow international negotiations to continue without being influenced by U.S. domestic action. The NRDC was persuaded that it was useful to keep the domestic and international processes in sync. The environmentalists also feared that pushing for regulations at that time might result in rules that were weak.

The delay was disheartening to Rowland although it failed to surprise him. After the poor response from other countries following the aerosol ban, Rowland had long doubted whether the United States would be willing to act alone on additional controls. But the country's real mistake was not in acting alone, he felt, but in not acting in a convincing enough manner. Rowland said in a March interview:

One of the saddest things about the regulation put in (on aerosols) was the response of our industry. . . . They went out and found other substitute uses that also release CFCs into the atmosphere. The Europeans' reaction was, "What kind of a ban is that?" The use of polyurethane foam has ballooned since that time. The industry is asking for more time in the guise of asking for international cooperation. The United States alone cannot solve the problem. But my recommendation is that we should go ahead anyway but push very hard for international co-operation.

Even without formal U.S. regulations, the chemical industry was under increasing pressure to reduce CFC emissions by the summer of 1987. Thanks to environmental groups and a few hard-working legislators, some CFC users were convinced to act before regulations were issued.

Ozone protection had become the number-one priority of environmental groups around the world. Even though it was willing to cut the EPA some slack, NRDC refused to relent in its campaign against ozone-depleting chemicals. And, while Doniger and his colleagues monitored the international proceedings, other environmental groups jumped on the bandwagon with various projects to help focus public attention on the issue.

Economist Daniel J. Dudek of the Environmental Defense Fund provided a study on the cost of reducing ozone depletion. While a phase-out would cost between $41 and $540 million annually, depending on how rapidly it was carried out, the damage resulting from increased ultraviolet light and global warming would also have its costs. Dudek estimated that ozone depletion over the next 40 years would result in 1.44 million cases of cancer, a 3 percent loss in cotton, and 1 percent loss in wheat and corn crops. Increased UV light would also cause paint to fade faster and materials like polymers to crack, resulting in repair costs of $10 to $27 million.

At the World Resources Institute and Worldwatch Institute, studies were completed to alert Americans to the effects of various ozone-control policies. The Environmental Defense Fund, Friends of the Earth, and Sierra Club initiated public education campaigns and began pressuring industry to own up to its responsibility.

The environmental campaigns were colorful, creative, and stinging in their attacks on industry. Friends of the Earth targeted fast-food restaurants for the "frivolous" use of CFCs in the making the styrofoam containers to keep hamburgers hot. (In 1987, Big Macs had become a kind of symbol for CFC pollution.) As part of its "Stratospheric Defense Initiative," Friends of the Earth began a "Styro Wars" campaign to lobby the Foodservice Packaging Institute. The British arm of the environmental group began a boycott of the 20 best-known CFC-aerosol products in England. And in Canada, the group sent aloft 1,000 helium-filled balloons on Parliament Hill to focus attention on its ozone protection campaign. The balloons symbolized the flight of

CFCs and halons and illustrated the global nature of the problem; like CFCs, balloons released in one city can be found later in another. Friends of the Earth also began a campaign to educate American hospital administrators about the use of CFCs in the sterilization of surgical equipment. This use accounted for only 3 percent of the U.S. consumption of CFCs, but the environmentalists pointed out, non-CFC alternative sterilizers were available.

On Capitol Hill, Senator Robert Stafford targeted high-profile CFC users in order to gain momentum for CFC reductions and call attention to products that used the chemicals. In February 1987, despite snickers that his attempts would fail, Stafford wrote to McDonald's chairman and CEO Fred Turner to suggest his franchises discontinue using foam food containers made with CFCs.

"It is unlikely that you have heard of this issue in the context of your business operations," Stafford wrote in a letter to Turner. "But I assure you, McDonald's is well known in the CFC debate."

Poor McDonald's. In 1976 it had switched from cardboard to styrofoam containers after commissioning a study from Stanford which showed that foam had less environmental impact than cardboard in terms of raw materials, water, and energy used in production.

But, noted Clifford Raber, the corporation's vice-president for government relations, "We realize, however, that we cannot stand still and must be prepared to respond to new information and scientific knowledge as it becomes available." The company announced it would review its packaging process.

In March, Stafford went after a smaller fry. He wrote to the Senate Rules Committee to suggest a study on eliminating containers made with CFCs in the Senate cafeteria. Stafford's suggestion was implemented by May with the House cafeteria following suit.

McDonald's still had not acted. But Stafford soon received some help from Sister Barbara Mary and her second-grade students at St. Catherine's School in Rialto, California. The students were lobbying their local McDonald's to drop its use of CFCs, Sister Barbara Mary informed Stafford in a letter. In July, Theresa Freeman of Vermonters Organized For Clean-up announced a statewide McDonald's Day of Action in Vermont on August 1—a boycott of the chain—to protect the use of CFCs.

On August 5, McDonald's waved the white flag. Raber said in a letter to Stafford that the company would switch packaging in all of its U.S. operations within 18 months. "While our decision will not have any realistic impact upon the level of CFCs being emitted we do believe it could be helpful in persuading others to make similar decisions."

CFCs used for food packaging accounted for less than 2 percent of total U.S. consumption but, said Stafford aide Curtis Moore, the campaign had raised public consciousness.

"It hit people right where they lived," said Moore.

□ □ □

EPA officials insisted that the Antarctic ozone hole was not a factor in discussions on international or domestic CFC regulations. They felt that the 1986 NOZE was inconclusive on the role of the chemicals. Should regulations be enacted on the theory that CFCs could cause sudden, dramatic losses of ozone and science later showed the hole to be caused by a natural phenomenon, any treaty or legislation could collapse.

"The issue of the moment, of course, is the ozone hole, but any country in the world that makes a policy decision based on the hole is making an error. There's just too much we don't know about it," Bob Watson told *Atlantic Monthly*.

And, noted Senator Dave Durenberger at the January hearings on the CFC regulations:

> If it weren't for the hole in the ozone layer over the South Pole, I doubt that we would be gathered here today on this subject. And if it turns out that the ozone hole is the result of some phenomena other than CFC emissions, then I suspect that this issue will fall to the bottom of the agenda again and we won't be able to convince the Congress to take the time to pay attention until the middle of the next decade when the increase in the Earth's average temperature will have risen above the point of yearly variation.

How much difference did the hole make in bringing the CFC-ozone theory to attention again? To Sherry Rowland, it had made all the difference.

"The hole changed everything," he told the *Los Angeles Times*, bluntly. "It got the governments to believe there is a problem."

The ozone hole, said Doniger, had thrust the CFC issue into a state of emergency. If the hole were found to be linked to CFCs, he warned in *Discover* magazine, "It will make Chernobyl (the Soviet nuclear plant explosion) look like a trash fire at the county dump."

And, added Moore, the environmental committee staffer who had seen the ozone issue rise and fall and rise again:

> Without the hole, you could make a very good, very credible scientific argument that (global depletion) might just be natural variation. The only thing that blew that apart was the hole. That's what is scary about the hole. The hole is merely one example of the phenomenon that people have yet to recognize. And that is that by the time you see the bullet coming toward you, it's too late to dodge it.

A draft of the international ozone protocol was written in the summer of 1987 in preparation for the September meeting in Montreal. For the U.S. delegation, it was time for a final summary of the scientific evidence for ozone depletion as it stood midway through 1987.

The U.S. delegation's science advisers, Dan Albritton and Bob Watson, wished they'd had just a little more time. More than a hundred scientists were about to depart for Antarctica in one of the most important atmospheric missions ever. And, in six months to a year, the data from that expedition would be in. NASA's weighty Trends Panel report on global ozone losses would also be completed within a year. They would know so much more. But Lee Thomas and his team had to know where things stood now, not a year from now.

Albritton felt it was important for the negotiators to be reminded of the outstanding questions that remained. For one thing, said Albritton, no one knew if Donald Heath's satellite data showing 4 percent global ozone depletion over the last seven years was correct. Second, the Dobson ground-station readings showing ozone loss, such as the one Rowland had studied at Arosa, had yet to be confirmed. Third, the cause of the ozone hole in Antarctica was still in question.

If a protocol was enacted in September, Albritton said, it would be based on the belief—on the theoretical prediction originated by Rowland and Molina—that if chlorine continued to increase then, ultimately, there would be a substantial loss of ozone particularly in the winter and in the high latitudes. After 14 years, scientists could agree on that much.

But there was no observed ozone loss directly linked to CFCs that could be put on the table as justification for a stronger protocol.

The scientific basis for the protocol would be rooted in theory. Any remaining disagreements over the document, Albritton knew, would involve questions over fairness, politics, and economics—not science. Although he had, months earlier, questioned his participation in what seemed to be a futile effort at solving an international crisis, Albritton had come to view the international ozone talks as something of a watershed event. The ministers and scientists gathered in Montreal were laying the groundwork for dealing with other global environmental problems, such as acid rain and ocean pollution. They were serving a tough apprenticeship. Action on ozone depletion would be taken on the strength of the belief that environmental disaster will occur sometime in the distant future if steps were not taken to prevent it. Something would be done prior to the observation of the problem. This had never been accomplished internationally. But a successful ozone protocol, Albritton and U.S. negotiators knew, would pave the way for talks on a more profound issue: the greenhouse effect.

Delegates from 43 countries met in Montreal on September 14. The talks started early. While the majority of the delegates worked out minor details of the treaty, a dozen representatives from the major CFC-using countries formed a small "working group" to discuss the heart of the protocol. UNEP executive director Mostafa Tolba participated in the working group's discussions, and he sensed trouble before the group even broke for lunch.

The U.S. team had thrown the discussions into chaos. It was demanding that the countries representing 90 percent of all CFC consumption sign the protocol before it would become binding. This would guarantee that the United States would not have to begin cutbacks before other countries. But, under this stipulation, either the Soviet Union, with 10 percent of all CFC consumption, or Japan, with 11 percent, would have veto power over the protocol.

The haggling went on all day and late into the evening. Thomas had fervently hoped that a protocol would be achieved but began thinking that the prospects looked dimmed. The European Economic Community and others were angrily protesting the U.S. proposition. The negotiations were scheduled to conclude the next day at 10:00 A.M. Breaking up around midnight, there was a feeling of discouragement. Some delegates remarked angrily that there would be no protocol if the United States insisted on its 90 percent consumption clause.

The hallways outside the meeting room were jammed with observers, and word spread rapidly that the talks were on the verge of breakdown. Albritton was numb. He knew that, back home, environmentalists were furious with the U.S. team's sudden demand for a 90 percent consumption clause in the protocol. If the talks fell through, the United States—the country that had united everyone in Montreal—would be blamed. Albritton found it hard to believe that after so much effort and time, the talks could collapse at the final meeting.

"Could it really happen?" he asked a few veteran negotiators.

"It could," he was told. Members of the negotiating teams returned to their rooms that night preparing to be disappointed in the morning.

Thomas was unwilling to give up. As the weary negotiators broke, he sought out and found the EEC representative Laurens Brinkhorst and asked to speak with him privately. The two men talked until nearly 3:00 A.M. They were able to reach a couple of compromises—options they felt they could present to the group in the morning. They arranged to meet at 8:00 A.M. the next morning to walk to the meeting room together to make sure they still agreed on the options.

Neither man had a change of heart. The next day, the working group heard Thomas and Brinkhorst's new proposals. The proposals were accepted but there were still some problems to work out. Tensions mounted once again. Time was running short.

At 5:00 P.M., the group arrived at a consensus and took the final agreement to the full group for its approval. The Montreal Protocol was signed by 43 nations that day. The participants agreed to a freeze on consumption and production of CFCs at 1986 levels by 1990; a reduction of 20 percent by January 1, 1994; and an additional 30 percent cut by January 1, 1999.

Most of the major CFC producers accepted some compromises and took home some victories. By getting the 50 percent cutback by 1999, the Europeans won a postponement of reductions. The United States won a hard-fought freeze on halons by the mid-1990s. Third World countries, such as Brazil and India, received extensions on the protocol's deadlines which granted them 10-year growth periods for CFCs and halons. The Soviet Union also got an exemption to allow it to open some CFC production plants already planned or under construction. The United States backed down from its original 90 percent consumption clause and agreed that nations representing two-thirds of all world CFC consumption must ratify the treaty for it to take effect. But Thomas won on another point that seemed to make up for areas in which the United States had compromised: The protocol contained a clause that all nations would meet again and would, ultimately, eliminate ozone-destroying chemicals.

To Thomas, the clause calling for follow-up negotiations to tighten the protocol was insurance. If the ozone loss in Antarctica was shown to be linked to chemicals, this clause would become very significant.

"(The ozone hole) emphasized to me that we needed to incorporate into the protocol the ability to periodically review the protocol based on new scientific information," Thomas said later. "We set up a time frame for review with a clear understanding that we would tighten or loosen depending on what that new scientific information showed us."

Around the world, the agreement—the first ever of its kind—was praised for its symbolism and criticized for its shortcomings.

The negotiators had had the opportunity to make great strides in reducing the uses of dangerous chemicals but had settled for a "symbolic" treaty, some environmentalists explained. At the World Resources Institute, Rafe Pomerance called the treaty a good first step but said that 85 percent cutbacks in CFCs were needed within five years to stop ozone depletion. Under the present treaty, depletion would continue. In fact, some observers pointed out, with all the exemptions written into the protocol, the document would actually achieve only 35 percent reductions by 1999, not 50 percent.

For EPA staffer Jim Losey, who was part of the small group of agency employees who began the drive toward CFC regulations under Ruckelshaus, the protocol was a bit disappointing. Losey had just missed his first day of law school in order to see the agreement through. But, having a drink with a British counterpart before leaving Montreal, he was forced to agree that some of the European Community's actions made sense. A few years ago, the United States had considered forcing a vote to require all countries to ban CFCs in aerosol cans. In hindsight, an overall approach to CFC and halon reductions, even though it was a weak one, was far better than enacting a series of use-specific treaties: first aerosols, then air conditioners, then solvents, and on down the line.

"I'm glad in 1987 we ended up with the Montreal Protocol rather than an aerosol (protocol) in 1985," Losey told the British delegate, conceding that the comprehensive approach was better. "We could be another 10 years filling in the other 10 big uses. If we had to go back and do a protocol on air-conditioning units and do a protocol piece by piece, it would take forever."

But, the British delegate admitted to Losey, the United States had pushed the other countries into agreeing to greater cutbacks.

Overall, Losey thought, things had turned out pretty well. The important thing was that there were prospects in the future for tighter regulations.

Rowland, too, saw hope for the eventual ban of all ozone-depleting chemicals. And although it was the first agreement to reduce CFCs since the 1978 aerosol ban, there was little joy in his voice upon hearing of the protocol. After all, he pointed out, it had been 14 years since he and Molina had blown the whistle.

"I think the situation is more dangerous than most of the nations are acknowledging," he said. "I'm sure there will be a call for more (reductions) as soon as the ink is dry on this treaty."

14
Proof—But Not Enough
September 1987—March 1988

Every room was booked in the Cabo de Hornos Hotel in Punta Arenas, Chile, by mid-August. The 150 American scientists had taken over the pleasant, picturesque town of 100,000 people. It was the moment the scientists had been anticipating for a year. Many were fatigued from the arduous preparations for the trip, but the rush of adrenalin more than made up for it. They had come here to do a very important job.

"I believe this is probably the single most important Earth science project in a decade," Watson told *Chemical and Engineering News* days before the trip. "Scientists—at least some of us—were getting complacent, thinking that they had started to understand the stratosphere and what controls ozone. This phenomenon in Antarctica was absolutely unexpected, absolutely unpredicted. We don't know if it's chemistry, we don't know if it's dynamics."

Project manager Estelle Condon was exhausted from six months of seven-day work weeks, and so were many other people who had worked overtime for months. There had been so many last-minute tasks. Aircraft use for atmospheric science, in general, had been a low priority for NASA. Both the DC-8 and ER-2 were 1950s aircraft and there were no spares. The DC-8 had come back from the aircraft modification facility in May—five months late—and had to be outfitted with lab equipment in the frantic weeks before the mission. The taxiway at the Punta Arenas airfield had been completed during the first week of August but the bathrooms and parking lot still weren't done. Ames officials shipped four 30-foot containers of equipment to Chile and rented a C-141 transport plane to carry other last-minute equipment.

It was the largest deployment Ames had ever done—and on the shortest notice.

The experiment, which would run until September 30, would feature the DC-8 and ER-2 carrying a total of 21 instruments. The instruments would measure ozone and the concentration of certain chemical species. It would determine temperatures and the aerosol content of the atmosphere. Computer modelers were brought in to analyze the data as soon as it came in.

There was much to be done in just getting the aircraft off the ground, however. The ER-2 had a limited flying range which meant that, departing from Punta Arenas, it would be unable to get very far over Antarctica. It was crucial, however, that the pilots fly into the hole, where they could get good data. In order to determine the best flight path, Watson had arranged to have satellite data from Goddard transmitted to Punta Arenas each day. The task was easier said than done. Normally, satellite data wasn't processed for many months. The Antarctic Airborne Ozone Expedition needed the data in 24 hours. A Goddard team was organized to handle the assignment, and a special communications system was put in place to send the information.

Weather was another problem that had to be addressed. Forecasting weather conditions at the altitudes at which the ER-2 would fly was difficult. The best forecasting models were at a British center. After lengthy discussions, NASA arranged to have the British meteorologists telefax about 300 weather charts a day to Punta Arenas. The charts fell into the hands of meteorologists Peter Salter and Adrian Tuck; two men who, except for the ER-2 pilots, may have been under more pressure than anyone there.

It was up to Tuck and Salter to review the weather information and decide whether it was a good day to fly. Although the pilots made the final decision on whether to fly, Tuck and Salter knew the pilots would make every effort to carry out their instructions. They didn't want to be the ones to endanger the pilots' lives.

On the first day of operation in Punta Arenas, maintenance teams were at the airport well before daybreak to make sure the planes were ready to fly. The ER-2 was scheduled for a 9:30 A.M. departure, which meant the scientists had to get up around 4:00 A.M. to prepare the instruments. Tuck and Salter would have to be at work long before that to determine if it was a good day to fly. They wanted to avoid rousing the scientists and pilots at 4:00 A.M. and then deciding to scratch the flight.

The ER-2 pilots were more than a little skeptical. From the beginning, the four pilots had had serious doubts about what they could achieve on the mission and had only reluctantly accepted the assignment. They would be at the mercy of a single engine, without Antarctic survival gear and with no place to go if the engine quit. Frankly, the idea didn't suit them very well at all. But they had taken their time in reaching a decision, researching what the environment was like, what sort of navigational aids would be available to them, and what the weather conditions would be like. Eventually, the group concluded that if they ran the operation from a north-south direction, along the Palmer Peninsula, then the risk level was within reason. If the plane should go down along this route, there was at least a slight chance that they could land near some manned site. But, chief pilot Ron Williams noted, "If you have trouble with an airplane, very seldom do you have that choice."

The NASA scientists hadn't convinced the pilots that the mission would be fruitful. But NASA had put a lot of importance on the program and the pilots were hired to do what was requested of them. Williams told Watson the group would go but added, "We're flying into an unknown environment and unknown conditions. We will go down there and start out slow and see if we can do it or not. There is a possibility that we might encounter conditions that we can't handle and we won't be able to do it."

□   □   □

Williams, however, dismissed all his doubts as he prepared for the first flight attempt on August 17. He had learned long ago that once he stepped into the aircraft, he had to disregard the risks and concentrate on flying the machine. A no-nonsense, deliberate man, Williams was calm about the task before him. "If you worried about those other things you wouldn't go in the first place," he said.

Donning an orange pressure suit, Williams began inhaling pure oxygen an hour before takeoff in order to cleanse nitrogen from his body. The oxygen bath would prevent the bends, which was caused by a rapid loss of air pressure. None of the journey would be easy. In the hangar, temperatures were usually −10°. With the airport hanging off the tip of South America, the winds were brutal. But it was nothing compared to what he would experience during the seven-hour flight.

Atmospheric models were used to brief the pilots on what to

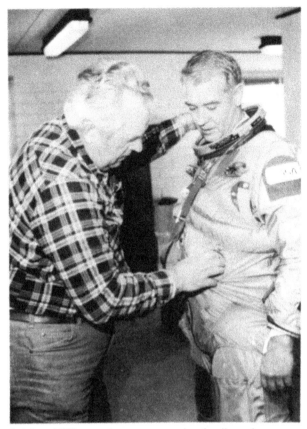

NASA pilot Ron Williams dons a pressurized suit and prepares to fly a high-altitude ER-2 aircraft into the ozone hole over Antarctica in September 1987. (Photograph courtesy of NASA)

expect over Antarctica. The winds at 60,000 feet were predicted to be in excess of 125 knots at times and the temperatures could be −90°. The pilots had little faith in the models. But, as Williams ascended southward he was astounded to see his plane slicing through an eerie veil comprised of tiny ice particles. At an altitude of 61,000 feet, he soared into the clouds and remained shrouded for the remainder of the flight. He had been flying the ER-2 for 20 years, but Williams had never encountered such conditions. He had never seen clouds above 60,000 feet. Except for very thin layers of clear sky, the clouds extended up to 85,000 feet. Well, he thought, the models were correct.

Along with the clouds, the former Vietnam War veteran had to contend with the sudden 150-knot winds that racked the plane. The

A NASA pilot climbs into the ER-2 prior to takeoff from Punta Arenas, Chile, in September 1987. The ER-2 is loaded with instruments to measure gases in the ozone hole over Antarctica. (Photograph courtesy of NASA)

final surprise came as the plane reached the edge of the stratosphere. At this stage, Williams knew that temperatures should be warmer. But that wasn't what was happening. Instead, he began to worry that his fuel lines would freeze. His instrument panel told him temperatures outside his plane were − 130°. The pilots had set a limit on the temperature they would risk flying in. And when Williams reached − 130° he turned the plane back, fearing his fuel would begin to freeze.

The plane had enough fuel to dip into the ozone hole, return to Punta Arenas, and continue another 600 miles north to Puerto Montt if they were unable to land at Punta Arenas. Landing the ER-2 was the riskiest part of the flight and drew a large crowd of nervous onlookers. Conditions for landing were marginal. With winds of 40 knots, the droopy wings of the ER-2 wobbled dangerously. Williams could afford to make two passes in an attempt to land. If he failed, he must fly to Puerto Montt where it would take five days to get the plane serviced and returned to its base—a time delay no one wanted. Luckily, the three pilots making the ER-2 flights never had to resort to their backup landing site.

Jim Anderson was doubly nervous as he watched the landing of the initial ER-2 flight. While hoping for a safe landing, he also wondered if his instrument had performed.

The instrument had worked well in test flights in California. But during those flights, temperatures in the troposphere were much warmer than the polar vortex over Antarctica, and there were questions about how the instrument would respond to the bitter cold. As Anderson had feared might happen, the instrument failed.

The problem was occurring somewhere in a maze of connectors in the device. Anderson's team had used only high-quality, gold-coated connectors hoping to avoid a failure at low temperatures. The initial data showed that the connector had worked until the pilot reached the vortex. When the temperatures warmed up, it began working, again.

"It was the most frightening kind of failure because it only occurs at the very low temperature end," Anderson explained. "There are hundreds of types of connectors like this on the instrument, and one of them was opening up."

NASA officials wanted to fly again the next day, and Anderson had less than 24 hours to diagnose and fix the instrument. The first task was to find out which connector had failed. That night, one member of Anderson's team stayed up, writing a computer program to locate the bad connector.

Anderson wasn't sure if the connector would fail again. But if it did, the new computer program should be able to pinpoint where the failure was occurring. The next day, with pilot Jim Barrilleaux in the cockpit, a review of the data showed that the failure reproduced itself and the computer had located it. The faulty piece was in the delicate communications line that connected the computer to the instrument. The connector was quickly replaced, and Anderson held his breath as the ER-2 made its third venture into the ozone hole. The pilots were endangering their lives to run an instrument that wasn't working. It was almost too much to bear. And so much was riding on his experiment.

When the ER-2 landed, Anderson rushed out to retrieve the computer data which was stored on a magnetic tape. The instrument had worked beautifully in this third flight, and Anderson was relieved. Now, maybe, they could get a peek at what might be causing the ozone depletion. What they got was much more than a peek.

Within a couple of hours Anderson was staring at the results of the chlorine monoxide measurements. Outside the ozone hole, chlorine monoxide levels were normal and low. But when the plane had hit the edge of the hole, the chemical had soared to levels almost 300 times what they had seen by balloons or other aircraft.

The instrument did not show any ozone depletion, but in the next three weeks, the scientists began to observe a coupling of ozone depletion and chlorine monoxide increases in what Anderson referred to as an anticorrelation. The fact that ozone decreased as chlorine monoxide increased was convincing evidence that chemicals were causing the depletion. The finding was repeated during the remainder of the 12 ER-2 flights. With the weather cooperating beautifully, the ER-2 pilots made many more flights than they had planned. And as their confidence grew, they began pushing deeper into the ozone hole and into colder temperatures.

By the time the expedition left Chile, there were few doubters. Anderson was known for being careful and meticulous. The results of the experiment were obvious. The only surprise was that it had come about so easily.

There was additional evidence to support Anderson's research. The DC-8 had carried out numerous successful experiments. Flying at an altitude of 42,000 feet, the DC-8 often hauled dozens of scientists, including Watson, during its 13 missions. While munching on popcorn during the 11-hour flights, the scientists were able to make important observations about the extent of ozone depletion, the surrounding weather conditions, and other key chemical species that might explain how and why the hole forms each year.

In order to avoid the confusion that had resulted from the 1986 NOZE press conference, Watson had arranged for the scientists to compile an end-of-mission press statement that would be released during the final days of work in Punta Arenas. The statement would only contain information that was clear and unambiguous. As it turned out, so much of the data was crystal clear that the statement contained almost everything the public needed to know. The public was told that almost half of the ozone layer over Antarctica disappeared during the months of August and September in 1987. And chlorine chemicals were causing the depletion.

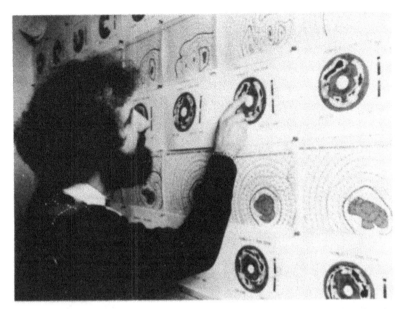

NASA scientist Robert Watson examines satellite data showing the location of the ozone hole. Satellite photographs were used to guide scientists studying ozone depletion during the 1987 Airborne Antarctic Ozone Experiment. (Photograph courtesy of NASA)

☐ ☐ ☐

Watson and Albritton wanted to move cautiously. Yes, the 1987 Airborne Antarctic Ozone Experiment had been a resounding success. Scientists had learned more about ozone depletion than they had thought possible. And it was clear that chemicals played a crucial role in the depletion. But, at a press conference at Goddard, the two senior scientists cautioned that more time and research were needed to assess the conditions in Antarctica and apply them to the rest of the world.

Within the inner circles of atmospheric scientists, the men and women who had worked so hard and argued so vehemently to explain the Antarctic phenomenon were nearing a consensus. At a conference in West Berlin during the first week of November, several of the proponents of the dynamics theories tried to salvage what they could of their arguments. But, one by one, the theories were dismissed. The evidence for chlorine was overwhelming. And even the chemists were stunned by Anderson's data which showed ozone dropping as chlorine

A NASA DC-8 aircraft served as a flying laboratory for scientists studying ozone depletion over Antarctica in September 1987. (Photograph courtesy of NASA)

monoxide increased. In particular, data gathered on September 16 showed that chlorine monoxide went up and ozone down about seven times in one hour. To illustrate the finding, Watson had brought a graph of the anticorrelation to the meeting.

The following week, at a meeting of the Ozone Trends Panel in Switzerland, the debate had come to an end. There was now detailed evidence of the presence of chlorine chemicals at work in the stratosphere. In fact, the dynamics theories had become the source of some joking. At one dinner conversation, Dan Albritton deadpanned to Mark Schoeberl, the formerly ardent supporter of the dynamics theories: "How are you coming with your completely dynamic explanation for acid rain?" Schoeberl didn't bat an eye. "I'm going to work on that as soon as I finish my chemical theory of thunderstorms."

Members of the airborne expedition weren't the only ones zeroing in on a chemical explanation for the ozone losses. In September, Farmer and his team of investigators published the results of their work during the first NOZE which showed a direct correlation between seasonal temperature changes and chlorine's transition from a non-

DC-8 mission manager George Grant at the control panel during a flight from Punta Arenas in September 1987. A DC-8 was also used to measure air samples over Antarctica. (Photograph courtesy of NASA)

reactive state to a reactive state. The researchers had also found high levels of chlorine nitrate, one of the two reservoirs for chlorine, outside the hole.

The second NOZE members had hauled 25,000 pounds of equipment to McMurdo and were able to confirm many of their 1986 findings as well. Dave Hofmann's group launched 65 balloons during the 1987 expedition and studied polar stratospheric clouds for the first time. And, although he had not gone on any of the Antarctic expeditions, Mario Molina was hard at work in his laboratory at JPL. After 15 years of work, Molina had the satisfaction of coming up with a key piece of evidence to prove the Rowland-Molina hypothesis.

In the November 27 issue of *Science*, Molina and his colleagues reported that they had successfully staged a series of reactions in the laboratory that could explain how ozone is lost from CFCs. The ex-

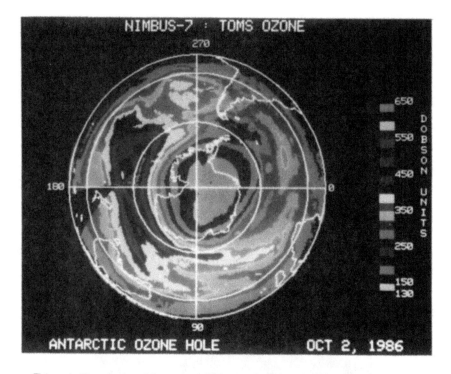

This satellite photo of the ozone hole on October 2, 1987, shows that the area of depletion is larger than the Antarctic continent. The satellite data were taken with the Total Ozone Mapping Spectrometer on NASA's Nimbus 7 satellite. (Photograph courtesy of NASA)

periment showed how chlorine is freed from hydrochloric acid, a reservoir, and converted to a form that breaks down with sunlight.

In the experiment, Molina created a thin film of ice in a narrow glass tube and then blew chlorine compounds into the tube to see how the chemicals would react. The results of the experiment showed that hydrochloric acid and chlorine nitrate can react in ice particles (such as those found in polar stratospheric clouds), setting up conditions that allow for the release of chlorine radicals that can attack ozone. Molina also showed that the sequence of events in the chain reaction occurred rapidly, with the chlorine nitrate requiring only a few collisions with ice particle surfaces to react with hydrochloric acid. This was particularly important because most scientists thought these heterogeneous reactions required hundreds of thousands of collisions before the reaction would occur. In the same issue of *Science*, investigators at SRI International in Menlo Park, California, published similar

NASA atmospheric chemist Mario Molina in his office at the Jet Propulsion Laboratory in Pasadena. (Photograph courtesy of Tom Hinkle)

results using a different method. The two elegant papers later won a distinguished award from *Science* magazine as the top research papers published by the journal in 1987.

The research proved what Rowland and Molina had said from the start: Chlorine destroys ozone.

It had been a rough few years under the ozone hole. And although the sad fact was now apparent that humans were responsible for the loss of the ozone layer due to the use of chlorine chemicals, some deserving scientists couldn't help feeling satisfied that their work had yielded such conclusive results in so short a time. For Christmas, Jim Anderson sent both Molina and Rowland a large, neatly framed graph of the telltale anticorrelation. Both hung the graph on their office walls.

The evidence was in. Now it was time for the world's leaders to do something about it. And there were signs that the ozone issue was, at last, getting the attention it deserved. On December 1, Senator Albert Gore of Tennessee, a candidate for the Democratic presidential nomination, raised the ozone issue before millions of television viewers on a prime-time debate with his fellow candidates. Responding to another candidate's question about environmental issues, Gore said that ozone protection was a major issue that must be discussed with the Soviets.

"The problem has not been that we don't know what to talk about," said Gore. "The problem we've had is (that) America has not been a leader."

Under the terms of the Montreal Protocol, UNEP director Mostafa Tolba had the option of asking negotiators to reconvene before the next scheduled round of talks—in 1990—if one-third of the parties agreed that circumstances warranted an emergency review. Tolba had said in Montreal that he would lobby for such a meeting if the Antarctic loss was directly linked to CFC emissions. The treaty was less than two months old when Tolba was asked to make good on his promise.

Scientists around the world agreed that the evidence provided by the 1987 expeditions, along with Molina's work, had significantly changed their view of how bad the ozone problem was. At the time of the protocol, the negotiators had been told to ignore the Antarctic phenomenon because of the lack of evidence that manmade chemicals were directly to blame. Scientists can now say with certainty that the hole in the ozone layer is manmade.

But there are other reasons scientists and others feel that the protocol is inadequate. Many of their reasons involve the doubts scientists harbor over ozone-depletion predictions. While years of additional research remain, most scientists believe that the Antarctic phenomenon has implications for the entire world. Very cold conditions are needed to set off the chemical chain reaction that destroys ozone over Antarctic, and it is doubtful that the same processes could occur elsewhere in the world—with the exception of wintertime losses in the Arctic and extreme northern latitudes. But scientists still don't know how much the seasonal Antarctic loss affects overall ozone totals in the Southern Hemisphere. When the polar vortex breaks up each spring, the ozone-poor air in the hole may disperse throughout the Southern Hemisphere, diluting the entire ozone shield.

Scientists are also studying the question of whether surfaces other than ice particles might be used in the heterogeneous reactions that lead to rapid ozone losses. There has been speculation that dust particles from volcanoes could provide such surfaces.

Finally, the very fact that atmospheric models failed to predict the Antarctic hole raises doubts about the ability of the models to accurately predict global ozone losses. Since it is possible that global ozone losses could be more severe than currently predicted, many experts feel it would be prudent to strengthen the Montreal Protocol.

"We've never seen a drop of this size in history," said Donald Heath. "This is the frightening thing. When you start seeing large changes in ozone where the theory cannot account for it, then this raises the question of how good are the model predictions? What are the consequences 20 or 50 years from now if the theory is wrong now?"

Such possibilities alarmed politicians, environmentalists, and scientists in the fall of 1987. In an October hearing before a congressional committee Rowland said, "We need to act now and impose severe restrictions of CFC emissions immediately if we want to bring the chlorine concentrations in the atmosphere under control by early in the next century."

Wilkniss helped to dramatize the situation by telling the panel that the ozone loss in Antarctica was now so severe that he feared for the safety of the scientists traveling there. The increased ultraviolet light, Wilkniss said, threatens the health of all the people of the Southern Hemisphere.

Even scientists who had either been reluctant to get involved in the politics of ozone depletion or who had chosen to withhold judgment over the past 14 years of the debate now had no qualms about speaking out on political solutions. Michael McElroy, who had contributed years of work on the issue but who had been unwilling to demand that industry discontinue the use of CFCs and halons, said that the ozone hole had changed things entirely. In a 1988 interview, McElroy said:

My view (before the ozone hole) was that there were lots of gaps in our understanding of the stratosphere and that there was a danger in going too far in regulation with too little information about what was really going on in the stratosphere. . . . There is a credibility problem which would mean that we would lose significantly. So I tended to take a very conservative view on what was required. And what turned me around was the indication that we had this very, very persistent and long-lived problem in Antarctica. And that it was not just an Antarctic problem, but it had implications for everywhere else. So that really changed my view of it. I was completely convinced at that point that it was time to be very serious about regulation.

McElroy continues to be a leader in urging bans on the use of ozone-depleting halon chemicals, such as those used in fire extinguishers. And while many scientists believe the Antarctic ozone hole cannot get much worse than it is, McElroy suggests that the depletion could accelerate if the Montreal Protocol isn't strengthened in the near future. At the October hearings, McElroy said the decision at Montreal not to consider the Antarctic depletion was inappropriate. "The situation has now changed," he said. "The chemistry responsible for the enormous drop in ozone over Antarctica is distinct."

The situation now, he said, "requires an appropriately radical response."

Shortly after the October hearings, which were called to review the Antarctic evidence, six U.S. Senators—John Chafee, Robert Stafford, Max Baucus, Dave Durenberger, George Mitchell, and Quentin Burdick—wrote Tolba urging him to reconvene negotiations in order to strengthen the protocol. Tolba was now in a sticky position. If he made good on his word he would risk infuriating the more reluctant signatories of the Montreal Protocol by telling them, less than two months after the meeting, that the historic agreement was suddenly no good and must be reworked.

The recent scientific evidence had cornered the EPA, too. Under the terms of the NRDC's lawsuit, the agency was scheduled to propose U.S. regulations on CFCs and halons on December 1. The administration had hoped to push along international and domestic regulations together and had succeeded in postponing the domestic regulations until after the Montreal Protocol. But now that strategy was backfiring. In the time that had elapsed between the protocol signing and the announcement of U.S. regulations, the new evidence had come in, making the protocol look weak. If the United States chose to ratify the protocol, as it was expected to do, it would be bound to honor the terms of the protocol. But there was no rule that it couldn't surpass those terms. Would the EPA choose to stay within the confines of the Montreal Protocol and ignore the new evidence?

The EPA's proposal, released in December, mirrored the terms of the protocol saying, in essence, that the protocol was adequate. In a lengthy report, the agency attempted to present the proposal as a breakthrough.

Without global controls, the world would lose half its ozone layer by 2075. But with the protocol, the loss would be less than 2 percent, the agency claimed. The EPA also presented a thorough list of the costs and benefits of controls. The costs included money lost during the transition period while switching from CFCs to substitutes, including temporary layoffs, new equipment, administrative costs, and the costs of unknown environmental hazards due to use of alternatives.

The benefits of regulations, however, included preventing increases in skin cancers and immune-system diseases. The EPA assessment predicted that without global controls, skin cancers would increase to 154.5 million by 2075 with 3.2 million deaths. With the Montreal Protocol, there would be 9.5 million cases with 142,000 deaths. Without the protocol, cataract cases could be expected to increase by 18 million, but with CFC reductions that number could be reduced by 92 percent.

The environmental benefits would be significant. Reducing CFCs would help slow the rise in global temperatures, beach erosion, loss of coastal wetlands, and sea-level rises due to the greenhouse effect. The reductions would also lessen the impact of ultraviolet light on marine ecosystems, crops, forests, and plants. The degradation of polymers could be prevented.

It sounded good. But, environmentalists pointed out, the assessment had failed to consider the new scientific evidence from Antarctica. It also lacked the most recent update on global ozone loss, which was due out from the Ozone Trends Panel in a matter of months. And finally, critics charged, the Montreal Protocol itself was misleading and was a poor model for the United States to base its strategy on. While the protocol called for eventual reductions of 50 percent, the exemptions given to Third World countries to allow for an increase in CFC use actually allowed for no more than a 35 percent global decrease of the chemicals. To protect the ozone layer and prevent accelerated losses, the protocol would have to achieve emission reductions of at least 85 percent. The U.S. regulations, Doniger said, did not meet the requirements of the Clean Air Act.

The U.S. action, however, relieved the chemical industry, which was already complaining about the economic costs of the Montreal Protocol. The regulations, said Alliance spokesman Kevin Fay, would cost industry $5.5 billion by 2010 and would double the cost of CFCs.

□   □   □

But industry was under pressure, too. Du Pont scientist Mack Mc-
Farland had participated in the airborne expedition and had firsthand
knowledge of the findings. Now it was up to McFarland to advise his
superiors at Du Pont's corporate headquarters of the rapidly changing
situation.

McFarland was, by all accounts, a nice guy. He was an atmospheric
scientist who had worked at the National Oceanic and Atmospheric
Administration in Boulder for several years and was, according to Molina, "one of us." A gregarious man, McFarland attended most of the
scientific conferences regarding ozone although he was often the only
industry representative present. He understood the science completely
and was not one to exploit the fact that there were still some uncertainties about what was happening in Antarctica. After the expedition
concluded, McFarland advised Du Pont officials that the hole was
caused by a combination of chemistry and meteorology.

At that point, Du Pont officials had accepted the fact that CFCs
were growing and that the current rate of growth warranted some
global controls. But the company's benevolence toward the environment stopped there. After hearing McFarland's summary, company
officials concluded, "that we didn't have enough information to separate the effects," said McFarland. Du Pont officials wanted to know
just how much of the depletion was due to meteorology and how
much was due to chemicals. Until they could be certain that ozone
depletion was due primarily to the chemicals, Du Pont officials would
resist a CFC phase-out.

In the offices of the Senate Committee on the Environment and
Public Works, however, a small group of senators was ready to take
on Du Pont for its stubborn refusal to face the facts. On February 22,
Stafford, Baucus, and Durenberger drafted a two-page letter to Richard
E. Heckert, the chairman of Du Pont. The senators packaged the letter
with newspaper clippings and advertisements published in the mid-
1970s which quoted Du Pont chairman Irving Shapiro as saying,
"Should reputable evidence show that some fluorocarbons cause a
health hazard through depletion of the ozone layer, we are prepared
to stop production of the offending compounds."

The senators challenged Heckert to live up to those words.

"We believe the time has arrived for the Du Pont Corporation to
fulfill that pledge. . . It is our judgment that there is no longer any
credible dispute that Freons do, and will continue to, damage human

health and inflict injury to the environment. We, therefore, respectfully request that the Du Pont Corporation commit itself to cease production and sale of all Freons that result in ozone depletion."

On March 4, the senators received their answer. Heckert wrote:

> Du Pont stands by its 1975 commitment to stop production of fully halogenated chlorofluorocarbons if their use poses a threat to health. This is consistent with Du Pont's long established policy that we will not produce a product unless it can be made, used, handled and disposed of safely and consistent with appropriate safety, health and environmental quality criteria.
>
> At the moment, scientific evidence does not point to the need for dramatic CFC emission reductions. There is no available measure of the contribution of CFCs to any observed ozone change. In fact, recent observations show a decrease in the amount of ultraviolet radiation from the sun reaching the United States.

Heckert's words stunned the senators and others who reviewed the letter. Heckert was obviously referring to a controversial study, published on February 12 in *Science*, that showed that ozone depletion was not contributing to an increase of ultraviolet radiation. The research paper, which was written by the National Cancer Institute and other contributors, had reviewed weather records in eight cities. The records showed that UV light had declined between 1974 and 1985. While the paper raised some legitimate questions, many critics charged that the decrease in ultraviolet light at ground level was probably due to an increase in pollutants in the urban areas where the study was made. An assessment of UV light in rural areas, where the air was much cleaner, might indeed show an increase of radiation. There were also questions about the accuracy of the instrument used to make the UV measurements.

Heckert concluded his letter by assuring the senators that Du Pont was concerned about ozone depletion and was looking at alternatives to CFCs. But, he said, "your proposal that Du Pont cease production of CFCs immediately is both unwarranted and counterproductive."

Less than three weeks later, Heckert was forced to change his mind.

15
The End of an Industry
March 1988—April 1988

Sherry Rowland was in Washington when his battle against chlorine chemicals and the people who make them began to draw to a close. During nine days in March 1988, Rowland and Molina gained the backing of the three factions they had long sought to convince: policy makers, scientists, and industry.

The demise of the American CFC industry began on March 14 when U.S. legislators unanimously approved ratification of the Montreal Protocol. The vote was 83 to 0. The United States was the second country, following Mexico, to approve the Montreal Protocol. But despite Congress' overwhelming support of the agreement, there was continuing criticism that the regulations did not achieve enough. Senator Chafee had threatened to introduce a resolution accompanying the ratification calling for additional regulations, but he failed to line up enough support in time for the roll call. Relieved, the Alliance for Responsible CFC Policy urged Reagan to sign the protocol as soon as possible, feeling that the terms of the protocol would be easier to live with than the ones Chafee was calling for.

Reagan, who had taken little public interest in the ozone issue during his eight years in office, signed the protocol on April 5 in Santa Barbara. It did not surprise anyone that the President, in praising efforts to protect the environment, noted that the protocol "creates incentives for new technologies." Reagan promised that the EPA would assess the treaty. "For our part, the United States will give the highest priority to analyzing and assessing the latest research findings to assure that the review process moves expeditiously."

□ □ □

It was probably a good thing for Bob Watson that the Senate had ratified the protocol. At 5:00 P.M. on the day the treaty was ratified, Watson ordered a copy of the Ozone Trends Panel Executive Summary to be delivered to the White House. In light of Congress' strong show of support for the Montreal Protocol, it is doubtful the administration would have tried to blunt the effects of the Trends Panel report—even though the report was the death blow to the CFC industry. But even if some people had wanted to deter Watson, the crafty administrator had given them little time to do so. The next morning, March 15, the massive Ozone Trends Panel report was released, stating that scientists, for the first time, had seen dramatic losses of ozone throughout the world.

Only two years after scientists had reported no observable ozone depletion anywhere, the Trends Panel announced that it had found losses in the Northern Hemisphere that were twice those predicted. According to the Trends Panel, the earth had already lost more ozone than the EPA predicted would occur under the Montreal Protocol by the year 2075. Instead of a 2 percent loss by 2075, as predicted by the EPA at the time of the protocol signing, the Trends Panel was reporting 1.7 to 3 percent ozone loss in the Northern Hemisphere in 1988. Watson reported that there was scientific consensus that manmade chemicals were responsible for the majority of ozone loss. The report certainly took some of the glow out of the previous day's ratification. Ratifying the Montreal Protocol as a means to end ozone depletion suddenly seemed about as effective as tossing a gallon of water on a five-acre brush fire.

According to the report, significant ozone losses had occurred in all latitudes between 30° north and 64° north—the bands that cover the United States, Canada, Western Europe, China, Japan, and the Soviet Union. The losses occurred mostly in the winter and were startling. While computer models suggested ozone should have changed from 1 to 2 percent, the panel found ozone losses in the winter of up to 6 percent. After allowing for the effects of natural variations, such as the solar cycle, the scientists found that ozone decreases ranged from 1.7 to 3 percent annually in the Northern Hemisphere and from 2.3 percent to 6 percent in the winter along those latitudes.

"This is the first report to say there are losses of ozone that cannot be ruled out by natural causes," said Rowland, who appeared at the

The Ozone Trends Panel report of March 1988 found that ozone losses are apparent in the Northern Hemisphere. Ozone depletion is more severe in the wintertime due to ice particles which are thought to make ozone-depleting chemicals more active.

Latitude band average	Average ozone loss	Winter average	Summer average
53–64°N	−2.3	−6.2	+0.4
40–52°N	−3.0	−4.7	−2.1
30–39°N	−1.7	−2.3	−1.9

The major findings of the 1988 Ozone Trends Panel show ozone losses over the Northern Hemisphere for the first time. The panel blamed the depletion on CFCs.

NASA press conference along with Watson and John Gille of the National Center for Atmospheric Research.

The difference in the panel's findings compared to earlier studies, said Rowland, was that this time the scientists looked at the data in small latitude bands on a month-by-month basis instead of lumping it together and averaging the data on an annual basis. The newer method, he explained, showed the wintertime losses that could be the result of cold-weather heterogeneous reactions such as those causing severe ozone losses in Antarctica.

While Molina had provided an important clue in explaining the Antarctic phenomenon, it was Rowland who helped show that global ozone levels were declining.

Watson had asked Rowland to chair the Trends Panel subcommittee assigned to study the world's Dobson stations. The choice of Rowland for this task was obvious. He was a senior scientist and, along with Irvine chemist Neil Harris, Rowland had already begun work looking at Dobson stations in the Northern Hemisphere. Rowland and Harris had, however, only looked at a few stations—Arosa, Caribou, and Bismarck—where they had seen signs of a greater change in ozone in the wintertime. But they couldn't draw any conclusions on the basis of three stations. For the Trends Panel analysis, the scientists were required to find the most reliable Dobson stations in the Northern Hemisphere and review their data. There were too few stations in the Southern Hemisphere to analyze ozone there.

To accomplish the tedious task, Rowland enlisted the help of Rumen Bojkov of the Atmospheric Environment Service in Canada. Bojkov is a young Bulgarian scientist with curly dark hair and an effervescent personality. He is also an expert at ozone measurements. Bojkov was already reviewing the accuracy of Dobson stations around the world when Rowland asked him to help with the Trends Panel report. Joining Rowland's committee, Bojkov began a thorough review of each station, sometimes applying small corrections to the data in order to achieve some consistency between the stations.

Gradually, Bojkov's work began to show a trend. The stations all seemed to agree: There were greater ozone losses in the wintertime.

While the Trends Panel had been organized to study Heath's satellite data, it was the Dobson station study that proved to be the key finding. Rowland's committee had looked at 18 Dobson stations and they all reported ozone levels going down. The ozone decreases could be attributed to the increasing use of trace gases such as CFCs, Watson said.

"The observed changes may be due wholly, or in part, to the increased atmospheric abundance of trace gases, primarily chlorofluorocarbons," the report stated. And, Watson explained at the press conference, "All of these gases are changing at a very rapid rate due to human activities."

According to the Trends Panel, the losses have serious implications for human health. More ultraviolet light is reaching the earth than at any time in earth's modern history, and pollution may be obscuring some of the radiation. For every 1 percent loss in ozone, there is a 2 percent increase in UV light, scientists estimate. Thus, a 2.5 ozone decrease year-round, such as that detected in some areas of the Northern Hemisphere, could lead to a 10 percent increase in skin cancer.

And, much to Heath's chagrin, the Trends Panel decided it could not use the satellite evidence. After struggling with the writing of his paper, Heath's research was finally published in *Nature* on March 17. But the Trends Panel had already chosen not to use the satellite data because of doubts about the instrument.

"Unfortunately, they suffer from instrumental degradation of the diffuser plate, the rate of which cannot be uniquely determined. Thus, the data archived as of 1987 cannot be used alone to derive reliable trends in global ozone," the report stated.

Besides the suspected accuracy of the satellite data, the data did not cover as long a period as the Dobsons, not even one solar cycle. But while Trends Panel members felt Heath's data could not be used alone, Heath's many years of work had demonstrated that total ozone can be measured by satellite. In pointing this out, the Trends Panel criticized the slow progress of the NOAA National Environmental Satellite and Data Information Service for failing, as of January 1988, to provide new data from the SBUV-2 instrument launched in 1984. For unknown reasons, that data had not yet been processed.

"There is urgent need for NOAA NESDIS to increase the priority given for the timely processing and validation of SBUV-2 ozone data if the United States is to have a viable national program for monitoring ozone," the Trends Panel charged.

But the report didn't stop with the analysis of the Dobson-versus-satellite data. While the first Trends Panel meeting had begun with the question, "Is Donald Heath right?", Watson had decided that if they were going to muster the resources and energy to assess global ozone then they ought to try and answer other questions. The widened scope included an assessment of the Antarctic phenomenon. With the 1987 expedition data now in hand, that part was easy.

The panel reported that, from 60° south to the pole, ozone had fallen to the lowest levels ever recorded during the previous spring. The hole had lingered longer than ever. And temperatures over Antarctica had stayed colder longer (well into December) than in previous years, indicating that PSCs lasted longer. The Trends Panel observed that, on a year-round basis, ozone had declined 5 percent at all latitudes south of 60° south. The findings severely implicated chlorine chemicals.

"This report says for the first time here is the corpse," said Gille.

Finally, the Trends Panel delivered a prediction for the future. Ironically, it found that, due to the changing solar cycle, ozone losses will not be as apparent in the near future. While the solar cycle bottomed out in 1985, making less ozone than is usual, the cycle is expected to peak in 1991, creating more ozone. However, after 1991, ozone losses due to trace gases will again be apparent. And, if emissions are not curbed, the losses can be expected to be greater than ever. With a CFC growth rate of 3 percent per year, along with current growth rates of other greenhouse gases, 10 percent of the earth's ozone layer will be destroyed in 70 years, the report stated.

The involvement of other trace gases will be an important factor in ozone depletion. An increase in methane, for example, reduces the rate of the chlorine catalytic destruction and creates ozone in the lower stratosphere. During the first week of March, Rowland and Donald Blake of the University of California, Irvine, reported in *Science* that

Concentrations of trace gases have been increasing in the atmosphere in the past decade. These gases absorb infrared radiation and contribute to the warming of the earth's atmosphere.

Gas	Observed trends, 1975–1985 (%)
Carbon dioxide	4.6
Methane	11.0
Nitrous oxide	3.5
CFC-11	103.0
CFC-12	101.0
Methylchloroform	155.0
Carbon tetrachloride	24.0

University of Chicago scientist Veerhabadrhan Ramanathan has shown that emissions of various gases contributing to greenhouse warming have increased in recent years. (Courtesy of V. Ramanathan, *Science Magazine*, vol. 240, pp. 293–299, 15 April 1988.)

methane has increased sharply since 1978. Methane comes from the decomposition of rice paddies and swamps, the intestines of cows, and wood digestion of termites. Human activities can increase methane because of raising more cows and cutting down forests, thus providing food for termites. While methane is a greenhouse gas and traps heat, Rowland and Blake reported that methane could have various effects regarding ozone. It could worsen the Antarctic ozone hole by helping to increase PSCs. (Methane in the stratosphere breaks down and releases hydrogen. Hydrogen eventually combines with oxygen to form water which can freeze in clouds.) Outside of the Antarctic, however, methane might react with chlorine to slow the rate of ozone depletion.

Overall, the Trends Panel concluded, such compensating factors, along with the change in the solar cycle, might offset ozone losses in the near future. But, after 1991, when the solar output begins to decline, ozone is expected to drop again from the increasing concentrations of trace gases.

The Ozone Trends Panel involved the world's leading experts: Watson, Albritton, Mahlman, Rowland, Schoeberl, Stolarski, Cicerone, Schmeltekopf, Johnson, and Wofsy, among others. The report, said Stolarski, who was cochairman, "was wonderful. I was surprised at how conclusive we were able to be about whether we believed it or didn't believe it."

And, at the press conference, reporters were equally impressed. After peppering Watson with questions, one longtime science writer observed that he had attended each of Watson's ozone-status summaries and that each time "you look a little worse and sound a little worse."

It was true. For atmospheric scientists, the past few years had been exciting and exhausting. But, said Gille to *The New York Times,* "For the first time we have a really definitive answer that ozone has decreased. We understand what is going on and we can predict it will be much more severe in the future."

The Ozone Trends Panel report was front-page news around the country and Watson, Rowland, and Gille made the evening's national newscasts. Reporters summarized the findings and applied them to the grim predictions about increases in skin cancer (made in previous NAS reports), telling Americans that the refrigerators that cool their soft drinks and the foam boxes that warm their hamburgers were to blame.

For a nation already overrun by a skin cancer epidemic, the news was frightening. Malignant melanoma has increased 93 percent in the past eight years. Basal and squamous cell cancers affect three in every ten people. Even without increased ultraviolet light, cases of skin cancer are predicted to double in 25 years due to life-style factors, such as spending more leisure time in the sun.

NAS studies showing that a 1 percent drop in ozone would increase skin cancers by 5 to 6 percent were similar to those reported by Dr. Margaret Kripke, an immunologist from the University of Texas Cancer Center who had chaired an EPA subcommittee report on the effects of UV light on human health, plants, and animals. But today, what concerns Kripke even more are the unknowns.

There is a great deal of data on the sun-skin cancer connection. But less is known about the potentially more damaging effects of UV light on the human immune system. According to preliminary studies, UV-B radiation damages immune cells in the skin called *Langerhans cells.* The damage, researchers believe, activates suppressor lymphocytes instead of activating protective immune responses. Suppressor cells circulate through the body causing a systemic immune weakening. Diseases like herpes and a tropical parasite disease are suspected to be affected by ultraviolet light. Epidemics triggered by increases in the sun's radiation seemed too horrible to imagine.

"We know there are immunological changes that occur in humans exposed to ultraviolet light," Kripke said. "What is important about that is the potential for that to influence the incidence of certain infectious diseases."

The Trends Panel report didn't sit very well with the participants of the American Society for Photobiology which was meeting in Colorado Springs when the report was released. Photobiologists are concerned with the effects of the sun's light on plants, fungi, crabs, and worms. The Trends Panel news had dropped on the annual conference like a bomb.

"This was quite a blow to us," admitted society president-elect Thomas Coohill. "We're very concerned about ozone depletion. Any depletion in the ozone layer means the genetic material starts absorbing more ultraviolet light, and it changes it. For evolution to occur, the genetic material has to be stable. Organisms don't have enough time to catch up from an evolutionary point of view."

While research on the effects of increased UV light on plants and animals is also scanty, some studies have shown that deleterious effects should be expected. Botanist Alan Temamura of the University of Maryland has tested 200 species of plants and found that UV-B causes cell and tissue damage in about two-thirds of them. And studies have shown that UV light could cause mutations in some microscopic marine organisms or changes in the eggs from larvae that float on the surface. Plankton could be very susceptible to change. Research by Donat Haber of the University of Marburg, West Germany, has shown that a 5 percent increase in UV light can cut the lifetime of some microorganisms by half.

The springtime ozone losses in Antarctica have provided a bizarre kind of laboratory for scientists to study the effects of increased ultraviolet light on marine organisms. Postdoctoral fellow Larry Weber and professor Sayed Z. El-Sayed of Texas A & M University have done several studies of the potential ecological effects of UV light on marine organisms and have found photoplankton was much less productive in enhanced UV light. Such changes in microscopic marine organisms could trigger changes throughout the Antarctic aquatic food chain, El-Sayed said.

Another study, by John Frederick and Hilary Snell of the University of Chicago, showed that even when ozone is depleted, the amount of UV light reaching Antarctica is not much more than that which reaches Miami in the summer. But plants and animals in Antarctica are not accustomed to Miami summers, and thus, cannot be expected to adapt well to increasing radiation.

The findings in Antarctica can be applied to the rest of the world should worldwide ozone levels continue to decline.

"Things like plankton seem to be at a level where a small change can have drastic effects," Coohill said. "If this (change) occurs on a large scale the effects reverberate throughout the food chain. What you would see is a lowering of crop yield. It would be subtle change. You wouldn't see crops dying in the field. It's an innocuous thing and the next thing you know, people are starving. In America, we don't worry so much about starvation. We worry mostly about the increased skin cancer. But when you look at it on a global scale you have to say that, more important than skin cancer, are the effects on plant life and animal life."

Of course, Coohill said, everyone hoped that the Trends Panel was somehow wrong. But that was doubtful, he added.

"The panel making the synopsis is a very high-level panel, and it's unlikely that it's wrong."

□   □   □

In a plush, ninth-floor suite of offices in Wilmington, Delaware, high-level officials for Du Pont were drawing the same conclusions about the high-level Trends Panel. Mack McFarland had attended the NASA press conference and, following the summary, walked with Du Pont environmental manager Joseph Steed and company spokeswoman Kathleen Forte to a telephone to call the company headquarters. The report was discussed briefly but, apparently, the significance of the findings hadn't dawned on the company's decision makers.

Returning to Wilmington, Steed and McFarland summarized the Trends Panel report and presented the evidence to top officials. They were beginning to become concerned.

Three days later, on March 18, Steed and McFarland were called to a meeting of the company's executive committee which included Heckert. McFarland explained that the Dobson stations had showed unrefutable signs of ozone losses. The ozone losses, said McFarland, were real. The meeting ended with Du Pont's decision to end the manufacture of ozone-depleting CFCs as soon as substitutes became available. For years, CFC industry officials had claimed that if chlorine chemicals did deplete ozone, the earth would deliver some kind of early warning. The Trends Panel finding of 1.7 to 3 percent ozone losses in the Northern Hemisphere was evidence that the early warning had come and gone.

While the decision has had enormous impact on the entire American CFC industry, Du Pont's move was made without emotion and cost it little. In 1987, CFCs accounted for less than 2 percent of the giant corporation's earnings and sales.

On the Saturday following Du Pont's decision, McFarland and the company's communications team began working on their presentation of the dramatic policy switch. By Wednesday, Steed had left for Washington to present the company's decision to government. On Thursday, Du Pont's customers would be notified and the announcement would be released to the public. In his office at the Jet Propulsion Laboratory, Mario Molina took a phone call from McFarland that Wednesday. McFarland wanted Molina to hear the news from him personally. Molina was appropriately stunned. He knew that Du Pont had, just a few weeks earlier, been using the National Cancer Institute paper on ultraviolet light to deny the need to act. Molina congratulated McFarland on the decision.

Du Pont, which supplies one-quarter of the world's CFCs, announced it would conduct a phase-out of the chemicals in an "orderly fashion." No timetable for the phase-out was given, and officials said the withdrawal of CFCs would depend on finding substitutes. But even without a timetable, it was clear that the end was in sight for Thomas Midgley's magic chemicals.

Du Pont's news shocked industry.

"It's hard to know quite how to react," a *Washington Post* editorial noted. "The new position is laudable; the old may turn out to have been disastrous."

In Washington, Du Pont officials visited Capitol Hill to explain the company's new position. The delegation did not drop in on Senator Robert Stafford, the person who had organized the earlier letter to Heckert demanding a CFC phase-out. The letter did not force the decision, Du Pont officials said. The Trends Panel's report on observable ozone losses was the only factor. "I hope it's clear that it's the science of the last week we're responding to," Steed said.

And, in a letter to Senators Baucus and Durenberger, Heckert explained:

In my letter of March 4, I noted that scientific evidence did not point to the need for dramatic CFC emission reductions. On March 15, we gained important new information from an international Ozone Trends Panel about the effects of fully halogenated chlorofluorocarbons on global ozone. . . . While we believe the short-term risks to health and the environment from CFCs is negligible, we nonetheless have concluded that additional actions should be taken for long-term protection of the ozone layer.

Stafford, who was to retire from the Senate in 1988 at the age of 75 and after 28 years in Congress, responded, "It is clear that the Du Pont Company, while still not ready to concede the magnitude of the danger, has nevertheless concluded that continued damage to the ozone layer by CFCs cannot be tolerated by our planet."

There was joy in the Senate environmental offices. One aide told the *Washington Post*, "This is the last nail in the coffin for CFCs. CFCs are finished."

But Du Pont's announcement, some suspected, also meant that the race for CFC substitutes was on, and the news had a profound ripple effect.

On April 13, makers of plastic foam food containers announced that they would convert to safe alternative products. With an estimated 30 to 50 percent of all food packaging made with CFCs, the move was significant because it represented the first major phase-out of CFCs by a major industry. The 15 companies involved vowed to convert to substitutes by the end of 1988, ending the use of CFCs in such products as fast-food containers and egg cartons. In this case, the switch to a CFC substitute would be relatively painless. Manufacturers reported that no costs would be passed on to consumers. And, even though CFCs in foam food containers represented only 3 percent of U.S. consumption of the chemicals, the decision was important for another reason. The action, Lee Thomas remarked, "demonstrates the degree of consensus in moving away from the use of CFCs."

No doubt Stafford's campaign and McDonald's earlier decision helped shape the consensus.

"Everybody . . . sees the importance this issue presents to the global ecology," said Jeffery Eves, vice-president of Fort Howard Corporation, the largest manufacturer of foam food containers. "More and more consumers have become knowledgeable. Every company that wants to stay in business wants to please its customers."

It was clearly becoming a liability to be linked to CFCs. The Society of the Plastic Industry, Inc., published a brochure stating that "molded foam cups do NOT contain CFCs." The organization stated that "a few organizations and local government bodies concerned with the effect of CFCs on the earth's stratospheric ozone layer have condemned *all* foam plastic containers in the belief they are all made with CFCs. This is not true."

To Rowland, Du Pont's action changed the entire situation. Although CFC manufacturers had always resisted regulations due to the expense and upheaval of finding substitutes, when the regulations appeared to be forthcoming it was to their advantage to make the switch quickly.

"The market is starting to break away," Rowland said. "People are looking for substitutes in order not to be involved. If you look at what happened when there was a discussion of banning aerosol propellants, then there was very strong economic motivation to be in with the substitutes. You certainly did not want to be the last one into the new market because you'd lose. To be the first one is a real advantage."

Business journals and newspapers followed the industry phase-out with headlines such as "The Rise and Fall of CFCs," and "The Decline of the CFC Empire." But industry officials disagreed over the pain the phase-out of CFCs would cause. Some predicted that, similar to the phase-out of CFCs in aerosols, a switch to alternatives wouldn't cause the plant closings, product failures, and higher prices that were feared. If manufacturers were allowed to phase out at a pace that was not disruptive to the economy and did not jeopardize safety, alternatives could be introduced with little turmoil.

"We think we can do it without undue disruption," Richard Barnett of the Alliance for Responsible CFC Policy told *The New York Times.* "There will be some anguish but nothing that will devastate society."

But others warned that substitutes would cost consumers more and that some appliances presently relying on CFCs might operate less efficiently with substitutes. Automobile air conditioners would have to be redesigned, experts warned. The thickness of steel refrigerator doors would have to be increased to make up for the decreased efficiency of the appliance due to the loss of CFCs.

The loss of CFCs 11 and 12 from the market "would completely alter the civilized world," Lawrence Woodward, a spokesman from Pennwalt Corporation told the *Christian Science Monitor.* "It's an unprecedented situation. An entire product line—the entire industry— is going to have to change."

Although few environmentalists would sympathize with industry's problem—after all, CFC manufacturers had had a 15-year warning and ample time to find safe alternatives but chose instead to invest their energies in fighting CFC regulations—the task of replacing CFCs in products and services worldwide will not be easy.

The uses of CFCs are staggering. In 1986, the last year in which statistics are available, some 2.5 billion pounds of CFCs were used worldwide. About 700 million pounds are consumed annually in the United States, about three pounds per person. According to manufacturers, 5,000 U.S. businesses (featuring 715,000 jobs) use goods and services linked to the chemicals.

But an EPA report released in May 1988 noted that progress in replacement of CFCs was already apparent.

"Companies are now beginning to reduce their use of CFCs and halons. Many are acting out of concern for the environment and to avoid price increases and chemical shortages when EPA's proposed regulation takes effect," the EPA report stated. The agency reported that, in most industries, widespread substitution of safer chemicals will require five to seven years of research and development.

Some of the progress is remarkable and will, as the president predicted, create economic advantages for creative and aggressive businesses. In January 1988, a small Fernandina Beach, Florida, company isolated an organic solvent from orange peels that has become the basis for a possible replacement for CFC-113—a solvent used to clean electronic components. The Florida company, Petroferm Inc., calls the chemical BIOACT EC-7 and has joined AT&T in testing it.

At Du Pont, the corporate giant has joined the Dolco Packaging Company to explore a chemical called Olefane to replace CFC-12 in the making of foam food packages. And in late 1988, Du Pont announced it expects to begin production of a new coolant by 1990 at a new $25 million plant in Corpus Christi, Texas.

As of late 1988, the progress on CFC substitutes was as follows:

◻ *Commercial and residential refrigeration and air conditioning*: The compound FC-134a is being studied as a replacement for CFC-12 but would cost approximately three to five times more. Another chemical, HCFC-22, is being tested for broader uses. HCFC-22 is made by adding a hydrogen molecule to CFCs which causes the compound to break apart in the lower atmosphere. While HCFC-22 is used in home air conditioners, using it in automobile air conditioners might require different tubes and compressors because it works at higher pressures than CFC-12.

□ *Mobile air conditioners*: FC-134a might also be used in place of CFC-12. HCFC-22 is another possible substitute.

□ *Production of plastic foam and foam insulation products*: HCFC-123 could replace CFC-11, the second most widely used CFC next to CFC-12. Another possible substitute is a chemical called HCFC-141b.

□ *Flexible polyurethane foam*: HCFC-141b and HCFC-123 are possible substitutes for CFC-11.

□ *Rigid polyurethane foam*: Again, the CFCs 11 and 12 used in this foam might be replaced by HCFC-123 or HCFC-141b. Rigid insulating foam, however, might be replaced with other types of insulation, such as those using fiberglass.

□ *Sterilizers*: CFC-12 might be replaced with pure ethylene oxide.

□ *Solvents*: BIOACT EC-7 is a possible replacement for CFC-113. The replacement of CFC-113 was a thorny point in the Montreal discussions because the chemical's use is rapidly rising, and because it is newer, less time has been available to consider substitutes. It may be the hardest CFC to replace. In February 1989, the Asahi Company, Ltd., announced that it had found a possible substitute called HCFC-225ca and HCFC-225cb. The new compounds have just one-tenth the ozone-depleting potential of CFC-113, but Asahi officials said it would take several more years to complete toxicity testing on the chemicals and to make them price-competitive.

□ *Halons*: The chemicals used in fire extinguishers and in some military applications will be harder to replace. However, the EPA is working with companies that use halons to adopt practices that will limit the emission of the chemicals.

In some cases, substitutes are not needed to reduce the use of CFCs. CFCs used as air-conditioner and refrigerator coolants can easily be removed before the appliance is trashed and the chemicals are released into the atmosphere. In February 1989, the EPA joined with the Mobile Air Conditioning Society in announcing a voluntary agreement to reduce emissions of the CFC refrigerant used in car and truck air conditioners by recycling the chemicals. According to some estimates, CFCs used in mobile air conditioning contribute up to 19 percent of CFC emissions. The Mobile Air Conditioning Society also revealed plans to phase out cans of CFC-12 for consumers who service their own vehicles in order to prevent individuals from venting CFCs into the atmosphere.

Environmentalists today are urging consumers to help phase out CFCs by replacing the air-conditioner hoses in their automobiles every three years (to prevent leaking of CFCs) and by asking their auto mechanics to drain the coolant into bottles rather than letting it evaporate. Consumers can also be discriminating in their shopping habits by refusing to purchase flexible foam products made with CFCs, by buying non-halon-using fire extinguishers and by asking their dry cleaners to avoid the use of CFC solvents in the cleaning of garments. The environmental group Friends of the Earth has alerted consumers to avoid the purchase of toys using CFCs such as Tricky Fun String, String Confetti, and other party toys that use CFCs to propel plastic streamers from spray cans.

Ironically, as the search for substitutes continued throughout 1988, it became apparent that CFC producers might reap billions of dollars in profits from the transition to alternatives. Production limitations like those called for by the EPA in its December proposal for regulations would give a few companies control over the limited supply, driving up the price of the chemicals and creating windfall profits for producers. Consumers, ultimately, will finance the search for substitutes by paying these higher prices on CFCs. But as the NRDC's Doniger noted in testimony in March, 1988, "There is no reason in the world why any portion of the industry—especially the companies responsible for damaging the ozone layer to start with—should reap a windfall profit from a phase-down."

The EPA has since announced that it would consider placing a regulatory fee on CFC manufacturers to tap into the profits they would get from controls on CFCs.

Rowland had heard the day before the announcement that Du Pont had something interesting to say. But, like everyone else, he learned of the company's decision through the media. After numerous go-arounds with industry on CFCs, little of what industry did or said surprised Rowland. But the kindly, tall man who had seemed to grow thinner and busier as the years went on, was stunned.

# #

But he was not speechless.

"They've always shown a remarkable ability not to believe what I've been saying," he told the *Baltimore Sun*. "Their decision is welcome now and would have been welcome anytime in the past 14 years."

Yet, Rowland was not expecting Du Pont to announce it would withdraw the chemicals prior to the government demanding a total phase-out.

"I had not anticipated that it would go from 'no' to 'yes' overnight," he said. "I suppose I sort of thought it would probably be based on regulation rather than on action before regulation."

It seemed strange that Rowland did not feel like celebrating Du Pont's decision or that dozens of colleagues didn't flood his office to congratulate him or that the major TV networks didn't call him and seek his response to the news. Rowland was not one to revel in others' misfortune, especially when this misfortune belonged to the entire planet. It had always been a shadow hanging over Rowland's career that being proven right meant the world would suffer environmental damage. Being wrong might have crippled his career but at least he wouldn't have to lay awake at night and worry that the world might not be fit for his grandchildren. It would have been nice if people had heeded his and Molina's warning back in 1974 and had corrected the problem. The Irvine scientists could have had the satisfaction of being right and having done something to prevent observable ozone depletion. But those days are gone. The damage has been done.

Still, Rowland said, "to see that there is a light at the end of the tunnel and that we actually might solve the problem in time to prevent the damage from being horrendous, that's very gratifying."

Throughout the spring of 1988, things did begin to change for Rowland. For almost 12 years he had received no offers to speak before industry groups, although such invitations are common among university chemists of Rowland's stature. The one exception had been a 1979 invitation to appear before a local dry cleaners' association in Los Angeles. CFCs are used in dry cleaning fluid and the businesspeople wanted to learn more about Rowland's theory. Even then, during a lull in the CFC-ozone argument, Du Pont dispatched three representatives to hear his talk.

What Rowland did and said had been a sore subject for industry for 15 years. And even toward the end of the debate, he was still considered the number-one enemy of the CFC industry. In April 1987, Rowland was asked to speak at an EPA-sponsored meeting for CFC industry officials on finding safe substitutes. But upon arriving at the meeting in California's Napa Valley, Rowland discovered that some CFC industry representatives were up in arms about his being there. They were refusing to discuss substitutes in Rowland's presence. EPA officials smothered the controversy by asking Rowland to speak in a less-obtrusive time slot during the meeting, and the Irvine chemist simply gave his speech and left. He wasn't terribly offended.

But Rowland was to benefit from his role in public education. In October 1987 he became the recipient of the prestigious Charles A. Dana Award for pioneering achievements in health. It was one of several national awards he has earned for his activities in the public interest. Finally, on February 8, 1989, the Science and Technology Foundation of Japan named Rowland as recipient of the 1989 Japan Prize. The prize, which carried an award of about $400,000, is one of the top international honors in scientific research. Rowland was nominated along with 200 other scientists in the category of environmental science and technology. Working tirelessly in his Irvine offices on methane emissions, Rowland was forced to put on a coat and tie and join his colleagues for a celebration. "It's hard contemplating the depletion of the ozone layer, but more difficult to determine and describe the meaning of awards," he said, uncomfortable under the glare of photographers' lights.

In perhaps the most appropriate of settings, Rowland's contributions to atmospheric science were addressed by Senator Dale Bumpers, a veteran of the ozone debate, during Senate subcommittee hearings in early April 1988 before Rowland testified for close to the fiftieth time.

"We would not even be aware of this problem had not scientists like Professor Rowland and others persevered even when the federal agencies and the Congress short-changed them in terms of research funds. I personally wish to thank Professor Rowland for his work and his patience," said Bumpers.

□ □ □

As the events of March transpired, reporters once again sought Rowland out for his knowledge. Invitations to speak came pouring in, too, including several requests to address major industry groups. While Rowland was happy to oblige, the past three years had taken enormous energy. His wife and daughter complained to him that they hadn't seen

Sherry Rowland in his office at the University of California, Irvine, in 1988. Rowland has been the recipient of many awards for his contributions to understanding atmospheric pollution. (Photograph courtesy of the University of California, Irvine)

such bags under his eyes since the early days of the ozone debate in 1974 and 1975. And Rowland admitted that he was tired of having to explain the issue.

"I sort of felt it was incumbent upon me to explain it," he said. "As we sometimes discussed, was it really going to be necessary that Mario and I explain it one-to-one to everybody in the world? At some point, if that were necessary, it was never going to work. At some point, people were going to have to learn it from some other source."

After having been a lonely voice for so long, at last that seemed to be happening.

16
The Importance of Knowing Sooner
May 1988—November 1988

Snowmass, Colorado, was a ghost town in May 1988, and one got the feeling that that was just how the organizers of the Polar Ozone Workshop wanted it. This was an important meeting and there was much work to be done.

While snow flurries fell outside and construction crews worked busily to prepare the resort for the upcoming summer season, scientists from around the world met in the village conference center to review the 1987 Antarctic expedition and to discuss a new problem: an ozone hole in the Arctic.

Little mystery remained about the cause of the Antarctic phenomenon. Most of the major evidence from the expeditions had come out much earlier. And there was an exchange of congratulations for the work that had been done.

"There is now an enormous database about Antarctica, perhaps more than anywhere in the world," Dan Albritton told the conference participants.

But many smaller questions concerning the cause of the ozone hole and its future remain. Predictions on the future of the ozone hole vary. One of the most frightening scenarios suggests that the ozone hole may feed on itself and become worse. Scientists say that the loss of ozone could make the area colder, thus allowing for increasing development of polar stratospheric clouds. PSCs, of course, provide the surfaces for the rapid chemical reactions that destroy ozone. The more ozone that is lost during one winter may create more virulent conditions the next year: even colder weather, even more clouds, and even more ozone destruction.

Others believe the hole may stay the same.

"As for what's going to happen in the future, most people are saying that it would be hard to lose more ozone than is lost there now," Albritton said. "So it will probably stay pretty much the same. It may last longer. It may have some small characteristics that would change. Most people are saying we're going to see pretty much the same ozone hole over the next few years."

But scientists face even more important questions regarding the presence of the awesome phenomenon, such as, will the ozone hole affect ozone levels in the entire Southern Hemisphere and can the same type of rapid depletion occur elsewhere?

Because of the polar vortex, the swirling mass of air surrounding the pole, it is most likely that the hole won't become larger. But, Albritton notes, there isn't enough evidence to say what happens in the spring when the hole breaks up. According to some reports, the breakup of the hole may dilute ozone throughout the Southern Hemisphere. In December 1987, after the breakup of the vortex, Melbourne, Australia, reported its lowest ozone measurement ever during an Australian summer.

Scientists also debate whether the heterogeneous chemistry that takes place in Antarctica can occur elsewhere in the world, causing other ozone holes. Because of the bitterly cold conditions in Antarctica and the unusual mountains and land masses that allow for the tightly contained polar vortex, many scientists doubt that such an ozone hole could form elsewhere.

"Given that no temperature is that low and the presence of polar stratospheric clouds, then it's very unlikely that anything equivalent could happen elsewhere," Albritton said. "It is very strongly linked to the particular meteorological conditions of Antarctica. It's not a cancer that's going to spread all over the world."

But, he added, "We do know that it is cold in the Arctic, though not as cold. We know that there is some degree of ice particles forming there. The question is could a small, scaled-down version be occurring there?"

That was the question most on the minds of the atmospheric scientists gathered in Snowmass. And it continues to be a question of great importance today. There are several reasons for the concern. For example, any depletion in the Arctic, however small, would have the potential of affecting many more people—the vast population of the upper Northern Hemisphere—than the Antarctic loss in the sparsely populated Southern Hemisphere.

And, Albritton notes, "We can't just say who cares about a 10-percent ozone hole. It would have the potential of doubling the ozone loss over those areas."

Currently, estimates of ozone losses in the Arctic range from 4 to 10 percent. But much of what is happening there is a mystery. Temperatures over the North Pole can drop to −80°C—cold enough to allow icy crystals of water and nitric acid to form. But the polar vortex is not as well defined in the Arctic and can form and break up several times during the course of the winter. Scientists do not know if the vortex can stick together long enough for the chemical conditions to occur that result in rapid ozone losses.

These are the questions that are currently under investigation. While a small team of ground-based investigators traveled to McMurdo again in 1988 in an attempt to find out more about the various types of polar stratospheric clouds there and mysteries of their formations, the focus of many researchers has shifted to the Arctic. In January 1988, a small team of scientists, including Susan Solomon, made a brief expedition to the Arctic from their base in Thule, Greenland. Solomon's luck held out. Joking that the polar vortex seems to follow her, the team had a clear sky to work with and the vortex lingered over Thule for two weeks. The scientists found evidence that chemicals may be at work in ozone losses there. Jim Anderson, who had made the key anticorrelation finding in Antarctica the previous September and a colleague at Harvard, Bill Brune, discovered elevated levels of chlorine monoxide and chlorine dioxide. In particular, the scientists found that chlorine monoxide was five to seven times higher at 61° north than at latitudes farther south—evidence that heterogeneous reactions may be at work.

In addition, a scientist for the Department of Environment in Canada announced that he had also found evidence of an Arctic ozone hole. The question, said scientist W.F.J. Evans, is "do we have a crack in the north end of the bathtub as well as the south? Even if the effect is smaller in the Northern Hemisphere, the effect is more important because that is where most of the people are."

The alarming findings resulted in plans for a major airborne expedition to the Arctic in January 1989 out of Stavanger, Norway. Estelle Condon, once again the project manager, said NASA's Arctic expedition was similar to the 1987 Antarctic mission. The ER-2 and DC-8 were used and carried some additional equipment. Many of the same scientists participated.

"We're trying to understand what is happening in these cold vortexes," Condon said. "Trying to understand whether the mechanisms we see at work in the Antarctic are in fact at work in the Arctic."

A crucial question is the processes that go on inside the polar stratospheric clouds. Once a mystery, scientists now know that there are several types of PSCs and that they are more pervasive than they first thought. Discovering the components that cause the clouds to form and in what temperatures is crucial to the ability of scientists to predict ozone depletion in the warmer Arctic.

The 1989 Arctic mission cost between $5 and $10 million. NASA's budget for stratospheric research is only $20 to $30 million a year. But, Watson said, "The question is can we afford to go. The answer is, in principle, no, but we have to. We have absolutely no choice, in my opinion. I believe it's an equally important aircraft campaign as the one going to Antarctica a year-and-a-half ago."

But the Arctic mission is already somewhat historic. In May, the United States and Soviet Union agreed to participate in the Arctic study together. While U.S. scientists worked out of Norway, Soviet scientists simultaneously conducted tests of the atmosphere over northern Siberia and the Arctic Ocean.

"This is a significant breakthrough," said Shelby Tilford of NASA's Earth Science and Applications Division. "This is the first time the Soviet Union has been involved in any type of coordinated ozone experiments."

In late February, scientists returned from the Arctic expedition with grim news. They announced that enough chlorine chemicals were found in the Arctic to cause an observed 5 percent reduction of ozone during the winter over much of the Northern Hemisphere. According to Watson, the same mechanisms that cause the ozone hole over Antarctica are taking place in the Arctic on a smaller scale. Unless the polar vortex breaks up soon, scientists predicted at a February 17 press conference, as much as 1 percent of the atmosphere's ozone could disappear each day for as long as 20 days as sunlight returns to the Arctic.

"We had no idea if the chemistry that occurs in Antarctica occurs in the Arctic, because Antarctica is much colder. Now we know that it does," Harvard's McElroy said at a press conference.

The findings should alarm members of Congress and provide new impetus to strengthen the terms of the Montreal Protocol, Watson added. Under the terms of the protocol, ozone-depleting chemicals in the atmosphere would double in the next 50 years.

"The incredible perturbation is a strong message to policy makers," he said of the chlorine-laden Arctic atmosphere.

If the Antarctic Ozone expeditions, the Ozone Trends Panel report, the Montreal Protocol, and the Du Pont decision failed to convince the world that ozone was an important issue, the heat wave of 1988 surely did.

In the summer of 1988, Americans sizzled in the worst drought in a century. While cities across the country reported dozens of days in 100-degree heat, Congress approved billions of dollars in drought aid to farmers. Taken alone, 1988 would have seemed like some kind of horrible fluke. But many scientists speculate that the weird weather patterns of the summer—including the development of one of the worst hurricanes on record—are linked to an overall pattern of global warming that may throw our world into chaos and crisis within a century.

The facts are startling: The world is more than 1 degree warmer than it was a century ago. This is far above the expected natural variation of 0.4 degree over a century. Last year was the warmest on record in the 100-year history of record keeping. Five of the warmest years have occurred during the 1980s. In testimony on June 24, 1988, NASA's James Hansen, one of the foremost researchers on global warming, blamed the current rapid warming trend on the greenhouse effect.

"It is time to stop waffling and say that the evidence is pretty strong that the greenhouse effect is here," Hansen said.

Hansen's research shows more warming in the winter than in the summer and more in the higher latitudes than near the equator. These observations closely match what computer models predict for warming.

"These are all expected signatures of the greenhouse effect," Hansen said.

The increasing emissions of carbon dioxide, CFCs, nitrous oxide, and methane along with ruination of the world's tropical rain forests, which help absorb harmful carbon dioxide excess, are expected to throw the climate into a state of precarious imbalance in the next century if these trends are not reversed, experts warn. Some scientists

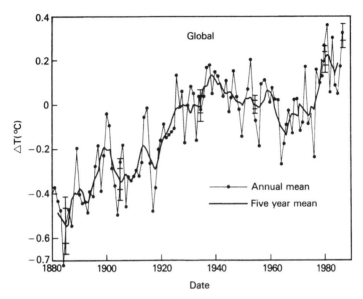

Studies show that global temperatures have increased during the past century with some increases occurring abruptly. Some scientists blamed the temperature increases on the buildup of greenhouse gases in the atmosphere. (Courtesy of Hansen and Lebedeff, *Science Magazine*, p. 240. 13 May 1988. Copyright by the American Geophysical Union.)

predict the world's temperatures may increase 3 to 9 degrees by the year 2050. Such a sudden deviation in global temperatures would not be unprecedented in earth's history, but the consequences of such warming are unknown in modern history, says Wallace S. Broecker of Columbia University's Lamont-Doherty Geophysical Institute. Studies of ice cores show that the climate has changed suddenly before.

"What these records indicate is that Earth's climate does not respond to forcing in a smooth and gradual way. Rather it responds in sharp jumps which involve large-scale reorganization of Earth's system," Broecker wrote in the July 1987 issue of *Nature*.

"We must consider the possibility that the main response of the system to our provocation of the atmosphere will come in jumps whose timing and magnitude are unpredictable."

The speed of greenhouse warming in the next century will be unprecedented, say experts.

"Mankind will be living in a world whose climate differs radically from anything in human memory," said Thomas Wigley, a climatologist from Great Britain.

How the world will cope with greenhouse warming is the center of much speculation. While computer models can predict some of the expected changes, much depends on what actions humans take to avert disaster and how rapidly those changes are made. One thing is certain: The release of manmade pollutants into the atmosphere has come about so fast and so furiously in the past century that the future of the earth can no longer be reliably determined. According to an April 1988 report by the World Climate Programme Impact Studies, which was founded by the World Meteorological Organization, records of climatic conditions over the past century can no longer serve as a guide to the future.

Scientists using computer models have attempted to predict what changes may be in store if such warming occurs. Many modelers agree that the greatest changes will occur in the high latitudes of the Northern Hemisphere. According to one description in *Science* magazine, the warming could cause Chicago's summers to become as warm as New Orleans' are now. Precipitation would increase in the winter and summer with the tropics becoming even wetter, while summer rains decrease in the mid-latitudes.

Scientists predict that for every increase of 1°C, climate zones will shift 60 to 90 miles north. While farmers could, conceivably, adjust to such shifts by moving or planting different crops, forests could not migrate fast enough to survive such rapid warming. The northern tundras, for example, would be swallowed by the sea.

The seas can be expected to swell from the melting of glaciers and the warming of the water. Increases of up to 4 feet are predicted by the middle of the next century. Beaches will be lost in many areas and coastal structures threatened. Fragile coastal wetlands will be submerged, affecting many migratory species. Flooding will increase. Ports will be damaged and water management systems will be in ruin.

The earth's inhabitants will experience great extremes in weather. It is this kind of sudden shift in climate that, some scientists believe, is already apparent and signals unnatural global warming.

The 1988 drought, for example, was caused by a stalled high-pressure system and a split jet stream. But scientists cannot explain what set up this unusual condition—other than the greenhouse effect. If the drought is one of the first signs of global warming, its implications are truly frightening. More than 40 percent of all counties in the United States were put on a drought disaster list in 1988, making them eligible for a variety of aid. The drought destroyed much of the Midwestern corn crop. It lowered the Mississippi River enough to halt traffic and allow some saltwater intrusion from the Gulf of Mexico. Massive forest fires roared in the West and brought into question whether the same environmental policies that have been used for decades to protect our natural resources will suffice in a world that is gradually heating up.

In September 1988, scientists testifying before a Senate committee even suggested that the devastating Hurricane Gilbert, which ripped through the Caribbean and the Gulf of Mexico in mid-September, might be an example of the consequences of global warming. The reason is simple, said Rowland, who was among the scientists who testified: The warming of ocean waters is known to trigger hurricanes. The warmer the water becomes, and the greater that pool of warm water spreads, the more hurricanes can be expected. Under the greenhouse effect, devastating storms might move into areas in which they were once rare.

Of course, there is a wide range of uncertainties to the greenhouse effect. The changes will vary from region to region. And much depends on the response of humans to the impending problem. The future emissions of fossil fuels, the rate of deforestation, and the strength of future treaties on CFC emissions will govern the eventual outcome. And scientists simply don't know how much the earth will change climatically for a given level of greenhouse warming.

"As you try to be more specific beyond the earth warming you get into the range of less and less certainty about the outcome," said Rafe Pomerance, a scientist with the World Resources Institute. "There are ways you can predict it. One is to look at past climates. Or look at computer models. It's better than guessing. And it's the best we've got. It remains to be determined how well those models work. We've only got one earth to experiment with."

According to Dr. Stephen Schneider, a senior scientist with the National Center for Atmospheric Research, small changes will occur first. But, he said, "It's going to be very hard to know if they're actually related to (the greenhouse effect). . . . The bottom line is that we insult the environment a lot faster than we understand the consequences. When you do that, you're guaranteed to have some nasty surprises."

How humans will respond to the greenhouse effect may depend, in a small but significant way, on future successes in protecting the ozone layer. International negotiations on ozone and CFC controls can contribute to solving the problem of global warming in two ways: One is to reduce CFCs and halons, which account for an estimated 25 percent of greenhouse gases. Second, the international negotiations on ozone can serve as a model for talks on the greenhouse effect.

The issues of ozone depletion and global warming, says EPA Administrator Lee Thomas, are "inexorably interconnected. . . . Both issues are clear examples of a 'global commons' environmental problem. All nations are responsible for contributing to recent changes in our atmosphere—although the industrially developed nations must shoulder most of the responsibility."

While the greenhouse effect is a much larger problem, it is similar to the ozone issue in an important way. Once again, policy makers are faced with an issue of scientific uncertainty whereby a "wait-and-see" policy might be disastrous but to act will entail large economic expenditures. As with the ozone issue, taking action on the greenhouse effect requires action to reduce future risks instead of imminent ones.

The Antarctic ozone hole, however, was a shocking reminder that the earth is capable of producing sudden surprises long before such consequences are expected, said Pomerance.

> There is no question that the Antarctic ozone hole has taught everyone a lesson. . . . You send the system in a particular direction and you get very surprising results, much worse than anyone predicted. . . . I think that lesson is being learned. We're very late (in acting) on the greenhouse effect because the 1988 atmosphere already has a huge warming built into it—on the order of probably one to two-and-a-half degrees or more. You have to act on the basis of prediction, not on confirmation. Because by the time you act on confirmation, you're too late.

Attempts at addressing the greenhouse effect began in earnest in 1988. In March, NASA announced plans to undertake the largest study of the earth as a single ecological system. And there are signs that the government is increasing its commitment to solving environmental problems. The EPA's budget for studying climate change and the stratosphere was $10 million in fiscal year 1986–87 but fell to $9 million the following year when the EPA's responsibilities were actually increased. The administration, however, proposed $17.7 million for research and regulatory analysis for 1988–89.

Internationally, there is progress, too. In June, scientists and policy makers met in Canada and discussed, for the first time, the need to reduce carbon dioxide emissions. Carbon dioxide emissions account for about half of all greenhouse gases. The result of an increasingly industrialized society, carbon dioxide levels have increased 25 percent this century. To coax the earth's climate back on a normal track, fossil fuel use must be cut by 60 percent, scientists say. Yet, consumption is predicted to double in the next 40 years if controls are not implemented.

Participants at the meeting in Canada agreed that they must address carbon dioxide controls and that a global convention is needed to do so. They set a goal of 20 percent reductions in emissions by industrialized countries by the end of the century.

"Carbon dioxide has been seen as such a daunting problem that nobody has said 'hey, we have to deal with this,' " said Pomerance. "This is the next target after CFCs."

According to the World Meteorological Organization report, four steps will be needed to respond to the greenhouse effect:

1. A protocol to protect the ozone layer.

2. Energy policies for increasing efficiency and reducing pollutants.

3. Policies to limit deforestation and begin reforestation.

4. Limit the growth of other trace gases such as nitrous oxide emissions produced by combustion and methane produced through agricultural processes.

But experts predict that an international agreement on the greenhouse effect will not be easy. Despite the fundamental agreement on what causes global warming, scientists disagree over the severity of its consequences and how readily living things can adapt. Alternatives to the fossil fuels that drive our economy are not handy and are fraught with problems of their own. The safe use of nuclear energy is one such example. Developing nations that are just beginning to reap the fruits of an industrialized society cannot be expected to reduce their reliance on fossil fuels at the demand of larger, richer countries.

Last July, 16 U.S. senators introduced a comprehensive package of legislation to combat the greenhouse effect, including proposals to develop safer nuclear power plants and to halt the destruction of forests. "The greenhouse effect is the most significant economic, political, environmental and human problem facing the 21st century," said Senator Tim Wirth of Colorado, the principal author of the legislation. Despite efforts to deal with the issues of ozone depletion, greenhouse effect, and acid rain comprehensively, however, those bills are being met with forceful opposition for the economic upheaval they would create.

As with the case of ozone depletion, will world leaders require such proof, that by the time it is accepted, dramatic changes will already have taken place? According to the EPA's John Hoffman, policy makers are beginning to realize the dangers of delaying action.

"It is not difficult to imagine that as recognition of the unique aspects of global atmospheric change becomes more widespread, decision makers will begin to focus on the effects of depletion and warming, recognizing the importance of knowing sooner," Hoffman said in August 1986.

The real question is whether our leaders recognize the importance of acting sooner and acting with an appropriate degree of prudence.

☐   ☐   ☐

On August 1, 1988, the Environmental Protection Agency ordered domestic regulations on CFCs that mirrored the Montreal Protocol.

The order was a disappointment to many. The Montreal Protocol was signed without the benefit of the 1987 Antarctic expedition findings or the Ozone Trends Panel report. The EPA had summaries of both findings and yet chose not to ask for deeper cuts of CFCs than the protocol instructed. Admitting that it might have underestimated the risk of ozone depletion, the EPA asked for public comment on the need for further regulations.

"After carefully considering the comments received, EPA has concluded that implementation of the Montreal Protocol is the best course the Agency can take at this time to securing adequate protection of stratospheric ozone," the agency stated. "Unilateral action by the United States would not significantly add to efforts to protect the ozone layer and could even be counterproductive by undermining other nations' incentive to participate in the Protocol."

Not surprisingly, the Natural Resources Defense Council disagreed. According to the NRDC, countries like Norway, Sweden, Canada, and West Germany have taken domestic actions that go beyond the protocol. And, the NRDC has charged that, in light of the new scientific evidence, the EPA order has failed to meet the requirements of the Clean Air Act.

Much to the credit of Lee Thomas, the EPA reversed itself less than two months later when Thomas announced that based on the Trends Panel report and a new EPA study the agency had, indeed, underestimated the expected degree of ozone depletion. "The depletion that has already occurred calls into question our earlier projections of future damage," Thomas said. "Regretfully, our new analysis predicts an even worse scenario than anticipated."

The EPA had maintained that 85 percent reductions would keep chlorine from growing in the atmosphere. But dramatic increases in CFC uses by Third World countries—which received exemptions under the Montreal Protocol allowing them to increase their consumption for several years—and the lack of worldwide restrictions on other ozone-depleting chemicals such as halons, methylchloroform, and carbon tetrachloride may negate the restrictions brought about by the protocol.

Methylchloroform and carbon tetrachloride, chemicals that are commonly used as fire extinguishers and as industrial solvents, could contribute to at least 10 percent of the chlorine in the atmosphere. According to a recent report by the Environmental Policy Institute in Washington, the increasing uses of these chemicals means that even under the Montreal Protocol the amount of chlorine in the atmosphere could almost quadruple in the next three decades.

Currently, the EPA estimates that chlorine in the stratosphere measures about 2.7 parts per billion. This amount could triple by 2075 if methylchloroform restrictions are not added to the present worldwide regulations.

The EPA estimates are, however, well under what Rowland predicts. According to Rowland's calculations, chlorine in the atmosphere would increase to 5.6 ppb by 2020 under the terms of the Montreal Protocol. Rowland's predictions include methylchloroform.

In testimony given in March 1988, Rowland made this prediction: "If a total phase-out of CFC emissions were accomplished globally over the next decade, then the chlorine concentrations would peak at about 4.4 ppb around the year 2000, and then slowly decline, reaching 4 ppb about the middle of the century." And, he added pointedly, "If the controls on CFCs announced in the United States in 1976 had been global and called for a total phase-out, the peak chlorine concentration would have been about 3.1 ppb."

Although the EPA's predictions are not as grim, Thomas has now called for a 100 percent reduction in the use of all CFCs and a freeze on methylchloroform by 1990. This action must be taken by all nations, with no exceptions, if chlorine in the stratosphere is to be stabilized and eventually reduced. According to the EPA, a complete phase-out of CFCs and halons is needed—along with a freeze on methylchloroform—for chlorine levels to stabilize for the next 100 years.

To be prudent, policy makers must act on the possible worst-case scenarios of ozone depletion because scientists can no longer reliably estimate future amounts of depletion. As Bob Watson reported during the Ozone Trends Panel press conference, "At the moment, it would not be unreasonable to say our models are not doing a good job to predict ozone changes. Therefore one would have to question whether we're underestimating the rate of change of ozone in the future."

And, said Rowland during testimony in September 1988, the fact that heterogeneous chemical reactions clearly play a role in ozone depletion makes future predictions fraught with uncertainties. Models, so far, cannot sufficiently reproduce these reactions.

"Although a strong consensus now exists that chlorine from the CFCs is actively destroying stratospheric ozone, predictions of future depletions of ozone by CFCs are as uncertain as they have ever been," he said.

"Our policy decisions about stratospheric ozone depletion now, and for a decade or more into the future, must be made without good quantitative guidelines of what the future holds for us."

□   □   □

How the world's leaders will deal with the new scientific evidence is open to speculation. An informal workshop to prepare for new negotiations was held in The Hague, Netherlands, in October in which scientists agreed that ozone levels were declining in the Northern Hemisphere. While the Antarctic ozone hole was smaller in 1988 than it was in 1987, scientists said climatological conditions were responsible and that the chlorine conditions that created the hole have not improved.

This scientific consensus should help pave the way for the next round of negotiations to strengthen the Montreal Protocol, said UNEP director Mustafa Tolba at a news conference following the Netherlands meeting. The protocol is expected to take effect sometime in 1989, following ratification by Japan and the Soviet Union. A meeting of parties will be held thereafter, followed by the completion of a scientific assessment in 1989 that will be used to determine whether the protocol should be amended. Representatives from the European Community and the United States have said they favored such a schedule.

"The Montreal Protocol calls for a review process and allows for new measures," Tolba said. "I believe we can speed up the process."

Many U.S. environmentalists have been heartened by this news. Others lament that it will take another one or two years before stricter measures can be put into writing.

"There is no longer any doubt about the inadequacy of the Montreal Protocol and the regulations proposed by EPA," said the NRDC's Doniger in testimony before a Senate subcommittee. "We've *already* suffered more ozone depletion than EPA predicted would occur under the Montreal agreement and the EPA regulations by the year 2050. In their present form they will not stop depletion, only slow its acceleration."

The situation is far worse than was predicted during the past year, Rafe Pomerance of the World Resources Institute told *The New York Times* in March 1988. "If the Montreal negotiators had had these findings in front of them they would have agreed to a total phase-out of CFCs."

While it is unlikely that the terms of the Montreal Protocol will remain intact much longer, the strength of a new treaty and the goal of worldwide cooperation are in serious doubt.

On March 2, 1989, at a meeting in Brussels, the environment ministers for the 12-nation European Community shocked the world by announcing they would agree to speed up CFC reductions, eliminating the compounds by the year 2000. The action was taken on the eve of a 100-nation Saving the Ozone Layer Conference in London and put pressure on the White House to support a stricter CFC phase-out. For the first time in the ozone crisis, the European countries had acted before the United States.

President George Bush responded swiftly on March 3 by calling for a ban by the end of the century if safe alternatives can be developed.

The actions by the European Community and American delegations raised the stakes at the London conference and demonstrated that it has become politically popular in the West to back ozone protection policies. Environmentalists warned against politically motivated remarks, however, and called for a hard deadline to end the use of ozone-depleting chemicals. At the NRDC, senior attorney David Wirth called Bush's qualifier of waiting for alternatives a "giant loophole that leaves too much of an out."

And Thomas Burke of the environmental group Green Alliance told the *Los Angeles Times*, "We're a long way from solving the problem of ozone depletion. . . . There's always a big gap between the rhetoric of government and the performance of government."

On March 5, the Soviet Union, China, and India soured the mood of the London conference by saying they opposed additional CFC cutbacks. China and India said that substitutes would be far too costly and called for a special fund to help Third World countries develop the technology to produce alternative chemicals. Some recent studies predict CFC substitutes will require new equipment—machinery that will be less durable and will consume more power. The more rapidly CFCs are replaced by substitutes, the more costly the process will be, representatives from China and India claimed.

While the Soviet Union had already agreed to a phase-out of 50 percent by the year 2000, it balked at a faster timetable, saying it was unconvinced of the severity of the problem.

"If there is scientific data that these substances should be phased out, we would support this idea," said Vladimir Zakharov, head of the Soviet delegation. He said the USSR did not accept western scientific findings of ozone losses in the Northern Hemisphere and a developing hole in the Arctic. The message seemed to directly contradict recent findings by Soviet scientists of an ozone "hole" over part of Europe.

Today, great concern remains that developing countries will expect industrialized countries to solve the ozone crisis, thereby defeating the only true solution to the problem: global effort. According to Irving Mintzer of the World Resources Institute, if China, India, Brazil, and Indonesia increased their CFC consumption to levels allowed by the Montreal Protocol, CFC production worldwide would double from the 1986 level.

Besides the resistance from developing countries, the temptation of Western leaders such as Thatcher and Bush to see substitutes on the market before outlawing CFCs may delay global action.

While some people argue that negotiators should reconvene immediately while the momentum is strong, other say a sudden call back to the table might upset the less enthusiastic participants of the protocol.

"The immediate thing is to get everyone to sign rather than go to people before they sign and say 'We've got to do something different,' because I'm afraid there could be the potential of upsetting the apple cart," said the U.S. negotiators' scientific adviser, Dan Albritton. "I would urge that people go to the corners of the world and try to convince people of the new science and urge them to sign. And then go back and say 'Look, now that we're all a club, what is the opinion on whether we should do something else.' "

Thomas agrees that the negotiations must move forward according to plan. While the continuous evolution of the scientific data could be used to ask for earlier talks, he worries that the scientific data could also be used to delay further negotiations or to even weaken the current treaty. The clause in the Montreal Protocol that demands additional negotiations to deal with evolving science is one of the most important contributions to the document. This kind of format—calling for periodic reassessments—will be crucial to progress on greenhouse effect negotiations, Thomas added.

"I think this is a real lesson . . . that will serve us well as we work on greenhouse," Thomas said.

When the nations do meet again, a momentous effort will be needed to obtain an agreement that will have truly lasting benefits for the ozone layer instead of merely symbolic value. Among the goals set forth by environmentalists: an entire phase-out of CFCs and halons by the end of the century. Anything less would be an admission that our society has failed to recognize "the importance of knowing sooner."

Tolba said in Vienna in March 1985, "If there is an environmental problem for which tardy response is absolutely unacceptable, it is the possible threat to the ozone layer. It is hard enough to cope with the permanent disappearance of a species, or the death of a lake, or the turning of fertile lands into desert. But in the case of ozone depletion, who could forgive us if we reacted too late?"

Americans are aware of the problem. A July 1988 poll released by the National Geographic Society showed that while 75 percent of Americans didn't know where the Persian Gulf was, 94 percent knew that damage to the ozone layer in one region could affect the entire world.

But to protect our ozone layer we must explore why we have become our own worst enemy. Nature is not an inexhaustible good. It is time to recognize that our stay on this planet cannot be rent free. If a total phase-out of chlorofluorocarbons were to occur tomorrow, it would be too late to stop the destruction from tons of CFCs already en route to ozone layer. We can't take them back. We will be reminded of their devastating effects well into the twenty-second century.

Bibliography

"Aerosol Link." *Time*, August 24, 1981.

"Aerosols' Latest Problem: Ozone." *Soap/Cosmetics/Chemical Specialties*, October 1974.

Alexander, George. "Increases in U.S. Skin Cancer Cases Seen." *Los Angeles Times*, October 4, 1974.

Auerbach, Stuart. "Banning of Aerosol Sprays as Hazardous to Ozone Proposed." *Washington Post*, June 13, 1975.

Beardsley, Tim. "U.S. Congress Says Climate Is Part of Politics." *Nature*, June 19, 1986.

Begley, Sharon. "The Silent Summer." *Newsweek*, June 23,1986.

Benchley, Bob. "Aerosol Debate Goes On." *Food and Drug Packaging*, June 3, 1976.

Boffey, Philip M. *The Brain Bank of America*, New York: McGraw-Hill, 1975.

Brand, David. "Is the Earth Warming Up?" *Time*, July 4, 1988.

Broeker, Wallace S. "Unpleasant Surprises in the Greenhouse?" *Nature*, July 9, 1987.

Brodeur, Paul. "Annals of Chemistry." *New Yorker*, April 7, 1975.

———. "Annals of Chemistry." *New Yorker*, June 9, 1986.

Brasseur, Guy. "The Endangered Ozone Layer." *Environment*, January-February 1987.

Bunker, Don L. "Ozone Alert: The Sky Is Limited." *The Nation*, November 30, 1974.

Burford, Anne M. *Are You Tough Enough*. New York: McGraw-Hill, 1986.

Chernow, Ron. "Should We Ban the Aerosol Bomb?" *Philadelphia Inquirer*, December 8, 1974.

Clark, William A. "What Are the Facts about Aerosol Sprays?" *Detroit News*, April 13, 1975.

Crawford, Mark. "EPA to Cut U.S. CFC Production to Protect Ozone in Stratosphere." *Science*, December 11, 1987.

————. "EPA: Ozone Treaty Weak." *Science*, October 7, 1988.

Davis, Donald A. "Ozone Depletion." *Drug and Cosmetic Industry*, April 1975.

————. "Aerosols." *Drug and Cosmetic Industry*, June 1975.

Dickson, David. "Congress Faces Decision on CFCs." *Nature*, September 3, 1981.

Doniger, David. "Politics of the Ozone Layer." *Issues in Science and Technology*, Spring 1988.

Dotto, Lydia, and Harold Schiff. *The Ozone War*. New York: Doubleday, 1978.

El-Sayed, Sayed Z. "Fragile Life Under the Ozone Hole." *Natural History*, October 1988.

"Engel Talks for Industry During National Packaging Week." *Aerosol Age*, December 1975.

Farman, J.C., et al. "Large Losses of Total Ozone in Antarctica Reveal CLOx/NOx Interaction." *Nature*, May 1985.

Fox, Jeffrey L. "Atmospheric Ozone Issue Looms Again." *Chemical & Engineering News*, October 15, 1979.

Frank, Peter H. "DuPont Takes Ozone Critics, Rivals By Surprise." *Baltimore Sun*, April 3, 1988.

Fritzsche, Doug. "Aerosol Ban Supported." *Daily Pilot*, November 25, 1974.

Ganz, Sol. "Letters." *Aerosol Age*, May 1975.

Gleick, James. "Hole in Ozone Over South Pole Worries Scientists." *The New York Times*, July 29, 1986.

————. "Even With Action Today, Ozone Loss Will Increase." *The New York Times*, March 20, 1988.

Hammond, Allen L. "Ozone Destruction: Problem's Scope Grows, It's Urgency Recedes." *Science*, March 28, 1975.

"Hysteria Ousts Science from Halocarbon Controversy." *New Scientist*, June 19, 1975.

Jenks, Richard L. "Ozone Odds." *Orange County Illustrated*, April 1976.

Kerr, Richard. "Antarctic Ozone Hole Is Still Deepening." *Science*, June 27, 1986.

————. "Taking Shots at Ozone Hole Theories." *Science*, November 14, 1986.

————. "The Weather in the Wake of El Nino." *Science*, May 13, 1988.

Kidder, Tracy. "Trouble in the Stratosphere." *The Atlantic*, November 1982.

Koenig, Richard. "Search for Substitutes for Refrigerants Shows Some Progress but Has Far to Go." *Wall Street Journal*, March 28, 1988.

Lemonick, Michael D. "The Heat Is On." *Time*, October 19, 1987.

"Letters." *Aerosol Age*, September 1975.

Lovelock, James. "Causes and Effects of Changes in Stratospheric Ozone: Update 1983." *Environment*, 26(10), p. 26.

Machta, Lester. "Ozone and Its Enemies." *NOAA Magazine*, January 1976.

Maugh, Thomas H. II. "Ozone Depletion Would Have Dire Effects." *Science*, January 25, 1980.

266	OZONE CRISIS

————. "Studies Renew Anxieties About Fading Ozone." *Los Angeles Times*, February 2, 1986.

————. "46 Nations Agree on Pact to Protect Ozone Layer." *Los Angeles Times*, September 17, 1987.

McCurdy, Patrick P. "Aerosol Makers Face Hard but Fair Shake." *Chemical Week*, June 11, 1975.

McElroy, Michael, et al. "Antarctic Ozone: Reductions Due to Synergistic Interactions of Chlorine and Bromine." *Nature*, June 19, 1986.

Mitchell, Sean. "The Politics of Freon." *The Nation*, June 28, 1975.

Molina, Mario, et al. "Antarctic Stratospheric Chemistry of Chlorine Nitrate, Hydrogen Chloride, and Ice: Release of Active Chlorine." *Science*, November 27, 1987.

Monastersky, Richard. "Decline of the CFC Empire." *Science News*, September 17, 1988.

"NAS Launches Study on Fluorocarbons." *Science News*, November 30, 1974.

"Nations Fail Global Move on CFC Problem." *Chemical & Engineering News*, January 25, 1982.

Norman, Colin. "Satellite Data Indicate Ozone Depletion." *Science*, September 4, 1981.

"Ozone Depletion: Early Evidence Comes In." *Science News*, August 22, 1981.

Pearson, Harry, "Trouble in the Ozone." *Newsday*, April 14, 1975.

Peterson, Cass. "'Greenhouse Effect' Needs More Study, U.S. Aide Says." *Washington Post*, June 12, 1986.

Raloff, J. "EPA Estimates Major Long-Term Ozone Risks." *Science News*, November 15, 1986.

Ramanathan, V. "The Greenhouse Theory of Climate Change: A Test by an Inadvertent Global Experiment." *Science*, April 15, 1988.

Rheem, Donald L. "Earth Atmosphere in More Danger than First Thought." *Christian Science Monitor*, June 12, 1986.

Rind, David. "The Greenhouse Effect: An Explanation." *EPA Journal*, December 1986.

Rogers, Michael. "Spray Away Layers of Ugly Ozone Fast, Fast, Fast!" *Rolling Stone*, December 4, 1975.

Ronberg, Cary, and Jerome Curry. "Trying to Stop Aerosol Spray." *St. Louis Post-Dispatch*, April 27, 1975.

Sand, Peter H. "The Vienna Convention Is Adopted." *Environment*, June 1985.

San Giovanni, Michael. "Exonerating Fluorocarbons." *Aerosol Age*, June 1976.

Schelling, Thomas C. "Anticipating Climate Change." *Environment*, December 1984.

Schiff, Harold. "Report on Reports." *Environment*, September 1982.

Schoeberl, Mark, and Arlin Krueger. "Overview of the Antarctic Ozone Depletion Issue." *Geophysical Research Letters*, November 1986.

Scotto, Joseph, et al. "Biologically Effective Ultraviolet Radiation: Surface Measurements in the United States, 1974 to 1985." *Science*, February 12, 1988.

Shabecoff, Philip. "Aide Sees Need for Steps Against Climate Change." *The New York Times*, June 12, 1986.

———. "Dozens of Nations Approve Accord to Protect Ozone." *The New York Times*, September 17, 1987.

———. "Industries Succeeding in Challenge to Cut Gases That Harm Ozone Layer." *The New York Times*, March 21, 1988.

Shell, Ellen Ruppel. "Weather Versus Chemicals." *Atlantic Monthly*, May 1987.

Sherwood, Martin. "Aerosol Scare 'May Be Over.' " *The Observer*, May 16, 1976.

Singer, Angela. "Danger Theory in the Can." *Yorkshire Post*, December 18, 1974.

Slagle, Alton. "The King of Aerosols Leads Battle to Defend His Domain." *Detroit Free Press*, May 23, 1976.

Solomon, Susan, et al. "On the Depletion of Antarctic Ozone." *Nature*, June 19, 1986.

"Son of Aerosol." *Time*, May 23, 1977.

Stolarski, Richard S. "The Antarctic Ozone Hole." *Scientific American*, January 1988.

Sullivan, Scott. "Nature's Revenge." *Newsweek*, March 2, 1987.

Sullivan, Walter. "Studies Are Cited to Show That Effects of Fluorocarbons on Ozone Layer May Be Cut 'Nearly to Zero.' " *The New York Times*, May 13, 1976.

Taubes, Gary. "Made In the Shade? No Way." *Discover*, August 1987.

Tuck, A.F. "Depletion of Antarctic Ozone." *Nature*, June 19, 1986.

Vromen, Galina. "Scientists Say Ozone Problems in Northern Hemisphere." Reuters News Service, October 18, 1988.

Washburn, Pat. "Pretty Hair May Well Be the Grim Wave of the Future." *Rochester New York Times-Union*, October 13, 1975.

"We Fail to Block a Possible Fatal Peril." *Newsday*, June 22, 1986.

Weisburd, S. "Pole's Ozone Hole: Who NOZE?" *Science News*, October 25, 1986.

Willis, Henny. "Scientists on Both Sides of Aerosol Spray Argument." *Eugene Oregon Register-Guard*, May 20, 1975.

Zurer, Pamela. "Complex Mission Set to Probe Origins of Antarctica Ozone Hole." *Chemical & Engineering News*, August 17, 1987.

Zurer, Pamela, et al. "Tending the Global Commons." *Chemical & Engineering News*, November 24, 1986.

Index

Index

Printed in the USA
CPSIA information can be obtained
at www.ICGtesting.com
JSHW021319221024
72173JS00001B/6